Annals of Mathematics Studies

Number 86

3-MANIFOLDS

BY

JOHN HEMPEL

PRINCETON UNIVERSITY PRESS
AND
UNIVERSITY OF TOKYO PRESS

PRINCETON, NEW JERSEY
1976

Published in Japan exclusively by
University of Tokyo Press;
In other parts of the world by
Princeton University Press

Printed in the United States of America
by Princeton University Press, Princeton, New Jersey

Library of Congress Cataloging in Publication data will
be found on the last printed page of this book

PREFACE

The aim of this work is to provide a consistent and systematic treatment of the topological structure of 3-dimensional manifolds. Our ultimate goal would be to provide (as have been done in dimension two) a "list" containing exactly one 3-manifold from each homeomorphism class together with an effective procedure for determining where a "given" 3-manifold belongs in this list. While this problem remains far from solution, the period since Papakyriakopoulos' proofs of Dehn's lemma and the loop and sphere theorems has produced considerable progress toward a solution and we have attempted to provide an organized account of these developments. We have excluded two topics: knot theory – which is a subject in itself and for which there are several reference works ([10], [16], [75]) and a consideration of local problems (wild embeddings, etc.) which are nicely covered in [13].

A basic principle in n-manifold topology is that k-dimensional homotopy theoretic information translates nicely to topological information provided the codimension, $n-k$, is sufficiently large ($n-k \geq 3$). This, together with duality, allows one to concentrate problems into the middle dimension, $[n/2]$, provided that $n - [n/2] \geq 3$. Of course this condition fails for $n < 5$. Also for $n = 3$ the middle dimension, one, involves the fundamental group – the only nonabelian homotopy group. This may help explain why the techniques (and results) in 3-manifold theory differ from the general theory ($n \geq 5$) and why the algebraic invariants involved are almost entirely group theoretic. The theme of this work is the role of the fundamental group of a 3-manifold in determining its topological structure.

We assume the reader is familiar with the basic elements of algebraic topology (covering spaces, Poincaré duality, the Hurewicz isomorphism theorem, and related topics). We will also make use of some facts from piecewise linear (p.l.) topology and from combinatorial group theory. In Chapter 1 we give a summary of the p.l. topology (regular neighborhoods, general position, etc.) which we use. We have included proofs of several theorems from group theory. Many of these, involving the structure of subgroups or quotient groups of a given group, translate (via covering spaces) to topological theorems and we have given topological proofs. We suggest that the reader may find it profitable to combine a study of 3-manifold topology with a study of combinatorial group theory.

We have placed exercises at appropriate points throughout the text. We have implicitly included many others by leaving details to be supplied by the reader.

Most of the material covered has appeared elsewhere in some form. We have made an effort to extend to nonorientable and/or bounded manifolds results which were previously known only for orientable and/or closed manifolds. We feel that we have achieved some economy in presentation by permuting the historical order of development. In particular, we have introduced the concept of "incompressible surface" as early as possible. Incompressible surfaces have turned out to be highly representative of the manifolds containing them. Combined with the tools provided by the loop and sphere theorems an analysis of the incompressible surfaces in a 3-manifold has proved to be the most effective approach to understanding the structure of the manifold. The most dramatic evidence of this is given in Chapter 13 where these ideas are used to show that a large class of 3-manifolds are completely determined by their fundamental group systems.

We have included a list of references which we hope will give proper credit to the original sources of the key ideas in the subject and will provide sufficient leads to further study. We have not attempted to provide a complete list of related works and apologize for all omissions.

This work developed through courses given at Rice University. I wish to express my gratitude to my students and colleagues for the stimulus they have given me.

My sincere appreciation goes to Sally Fanning, Janet Gordon, and Wanna King for their dedicated and able work in preparing the manuscript and to the National Science Foundation for their support.

JOHN HEMPEL

CONTENTS

3-Manifolds

CHAPTER I
PRELIMINARIES

We will approach the study of 3-manifolds from the piecewise linear (p.l.) point of view. This choice is prompted partly by tradition, partly by the ease with which low dimensional polyhedra may be visualized, and partly because of the technical convenience afforded by the "finiteness" of polyhedra. On the last point, while many of the standard arguments can be translated from the p.l. theory to the differentiable theory by replacing general position by transversality, there are some notable exceptions, e.g. proofs of the loop theorem and the existence of hierarchies, which do not seem well suited to differentiable techniques. Since each topological 3-manifold has a p.l. structure unique up to p.l. homeomorphism [71], [6] and a differentiable structure unique up to diffeomorphism [73], [112], our choice causes no loss of generality – theorems in the p.l. setting have direct analogues in the differentiable setting (as well as in the locally flat topological setting). Our choice does prohibit consideration of "wild" (nonlocally flat) embeddings of submanifolds, and we do not consider such matters. Thus we work entirely within the p.l. category; from Chapter 2 on the prefix p.l. will be understood to be attached to the terms manifold, submanifold, map, etc., unless otherwise indicated.

We will assume some basic elements from algebraic topology (e.g. Poincaré duality, Hurewicz Theorem, etc.), from group theory (most of which can be found in [63]), and from p.l. topology. For completeness we state, in the remainder of this chapter, the facts from p.l. topology one needs to begin with, and refer to [29], [47], [88], or [114] for a systematic development of the theory.

Definitions

We will denote n-dimensional Euclidean space by R^n, the unit ball $\{x \in R^n : \|x\| \leq 1\}$ by B^n, and the unit sphere $\{x \in R^n : \|x\| = 1\}$ by S^{n-1} and will call a space homeomorphic to $B^n(S^{n-1})$ an n-*cell* ((n–1)-*sphere*).

A (topological) n-*manifold* is a separable metric space each of whose points has an open neighborhood homeomorphic to either R^n or to $R^n_+ = \{x \in R^n : x_n \geq 0\}$. The *boundary* of an n-manifold M, denoted ∂M, is the set of points of M having neighborhoods homeomorphic to R^n_+; the *interior* of M, denoted Int M, is $M - \partial M$. By invariance of domain ∂M is either empty or an (n–1)-manifold and $\partial\partial M = \emptyset$. A manifold is *closed* if it is compact and has empty boundary and is *open* if it has no compact component and has empty boundary.

We will view a *simplicial complex* as a locally finite collection, K, of (closed) simplexes in some R^n satisfying

(i) If $\sigma \in K$ and τ is a face of σ, then $\tau \in K$.

(ii) If $\sigma, \tau \in K$, then $\sigma \cap \tau$ is a face of both σ and of τ.

We will denote the underlying space of K by $|K| = \cup\{\sigma : \sigma \in K\}$. By a *subdivision* of K we mean a simplicial complex L such that $|L| = |K|$ (as sets) and each simplex of L lies in some simplex of K. For simplicial complexes K_1, K_2 a map $f: |K_1| \to |K_2|$ is *piecewise linear* provided there exist subdivisions L_1 of K_1 and L_2 of K_2 with respect to which f is simplicial i.e. f takes vertices of L_1 to vertices of L_2 and takes each simplex of L_1 linearly (in terms of barycentric coordinates) onto a simplex of L_2. It is an elementary, but not altogether trivial, fact that the composition of piecewise linear maps is piecewise linear.

A *triangulation* of a space X is a pair (T, h) where T is a simplicial complex and $h: |T| \to X$ is a homeomorphism. Two triangulations (T_1, h_1) and (T_2, h_2) are *compatible* provided $h_2^{-1}h_1 : |T_1| \to |T_2|$ is piecewise linear.

For K a simplicial complex and σ a simplex of K, the star of σ with respect to K, $st(\sigma, K)$, is the subcomplex of K consisting of all simplexes of K which meet σ together with all their faces, and the *link* of σ with respect to K, $\mathrm{lk}(\sigma,K)$, is the subcomplex consisting of all simplexes of K which do not meet σ but which are faces of some simplex of K containing σ.

A triangulation (T,h) of an n-manifold, M, is *combinatorial* provided that for each vertex v of T, $|\mathrm{lk}(v,T)|$ is piecewise linearly homeomorphic to an (n−1)-simplex or the boundary of an n-simplex according as $h(v) \in \partial M$ or $h(v) \in \mathrm{Int}\, M$. This implies the more general fact that for each simplex σ of T, $|\mathrm{lk}(\sigma,T)|$ is p.l. homeomorphic to an $(n-\dim\sigma-1)$-simplex or the boundary of an $(n-\dim\sigma)$-simplex according as $h(\sigma) \subset \partial M$ or $h(\sigma) \not\subset \partial M$. If (K,h) is a combinatorial triangulation of M and L is a subdivision of K then (L,h) is also a combinatorial triangulation of M (c.f. [2] or [77]). A *p.l. structure* on a manifold M is a maximal, non-empty collection of compatible combinatorial triangulations of M. By a *p.l. manifold* we will mean a manifold M together with a p.l. structure on M. A map $f: M_1 \to M_2$ between p.l. manifolds is a *p.l. map* provided that for some (hence any) triangulations (T_i,h_i) $i = 1,2$ of M_i in the associated p.l. structures $h_2^{-1} f h_1 : |T_1| \to |T_2|$ is piecewise linear. We note [55] that there exist manifolds with inequivalent p.l. structures (i.e. homeomorphic p.l. manifolds which are not p.l. homeomorphic) and there exist manifolds with no p.l. structure whatever. The possibility that such manifolds might still admit (non-combinatorial) triangulations is still open and has some interesting consequences (see [93]). As previously noted such difficulties do not arise in low dimensions: each manifold of dimension at most 3 has a p.l. structure unique up to p.l. homeomorphism.

A submanifold N of a p.l. manifold M is a *p.l. submanifold* if there is a triangulation (T,h) in the p.l. structure on M and a subcomplex S of T such that $(S, h||S|)$ is a combinatorial triangulation of N (and hence determines a p.l. structure on N). This definition allows local knotting (i.e. the pair $(|\mathrm{lk}(v,T)|, |\mathrm{lk}(v,S)|)$ need not be p.l. homeomorphic to the standard sphere or ball pair of appropriate dimensions); however, if dim $M \leq 3$, all p.l. submanifolds are locally unknotted. A submanifold N of M is *proper* in M if $N \cap \partial M = \partial N$.

By an *orientation* of a p.l. n-manifold, M, we will mean a consistent orientation of the n-simplexes in a triangulation T of M. Here an orientation of an n-simplex is an equivalence class, modulo even permutation, of orderings of its vertices. Using square brackets for equivalence classes and the convention $-[v_0, v_1, \cdots, v_n]$ to denote the opposite orientation (i.e. $[v_1, v_0, \cdots, v_n]$), the orientation on the (n−1)-face opposite v_i *induced* by the orientation $[v_0, \cdots, v_n]$ is defined to be $(-1)^i[v_0, \cdots, \hat{v}_i, \cdots, v_n]$. Thus an orientation of M is a choice of an orientation for each n-simplex of T such that if an (n−1)-simplex τ is a face of two n-simplexes σ_1 and σ_2 of T, then the orientation on τ induced from that on σ_1 is opposite to the one induced from σ_2. Clearly a compact, connected n-manifold is *orientable* if and only if $H_n(M, \partial M) = Z$ and an orientation of M corresponds to a choice of generator for Z. We use the terms *oriented manifold* to mean a manifold together with a choice of orientation for it, and *unoriented manifold* to mean one which has not been oriented (whether or not it is possible to do so), and *nonorientable manifold* to mean one which can't be oriented.

Basic Theorems

By a p.l. *n-cell* (p.l. *(n−1)-sphere*) we mean a p.l. manifold p.l. homeomorphic to an n-simplex (its boundary).

Many of the elementary theorems, including the following three can be found in the early works of J. W. Alexander [2], and M. H. A. Newman [76], [77] as well as the reference works mentioned earlier.

1.1. THEOREM. *If* M *is a p.l.* n*-sphere and* C *is a p.l. submanifold which is a p.l.* n*-cell, then* $\overline{M - C}$ *is a p.l. submanifold which is a p.l.* n*-cell.*

1.2. THEOREM. *If* C *is a p.l.* n*-cell, then any p.l. homeomorphism of* ∂C *to itself can be extended to a p.l. homeomorphism of* C *to itself.*

1.3. THEOREM. *If* M *is a p.l.* n*-manifold and* C *is a p.l.* n*-cell such that* $M \cap C = \partial M \cap \partial C$ *is a p.l.* (n–1)*-cell (as a p.l. submanifold of both* M *and of* C*), then* M *is p.l. homeomorphic to* $M \cup C$.

The next two theorems are due to V. K. A. M. Gugenheim [30].

1.4 THEOREM. *If* M *is a p.l.* n*-cell or a p.l.* n*-sphere, then any orientation preserving p.l. homeomorphism of* M *onto itself is p.l. isotopic to the identity.*

1.5 THEOREM. *If* M *is a p.l.* n*-manifold,* C_1 *and* C_2 *are p.l.* n*-cells (as p.l. submanifolds) in* Int M *and* X *is any closed subset of* M *such that* $C_1 \cup C_2$ *lies in a component of* $M - X$, *then there is a p.l. isotopy* $\phi : M \times I \to M$ *such that* $\phi_0 = 1$, $\phi_t | X = 1$ *for all* $t \in I$, *and* $\phi_1(C_1) = C_2$.

Regular Neighborhoods

The theory of regular neighborhoods was developed by J. H. C. Whitehead [109]. We describe the essential features.

If K is a simplicial complex, σ is a simplex of K and τ is a face of σ, with $\dim \tau = \dim \sigma - 1$, which is not a proper face of any other simplex of K, then the complex $K - \{\sigma, \tau\}$ is said to be obtained from K by an *elementary collapsing*. If a subcomplex L of K is obtained from K by a finite sequence of elementary collapsings, we say K collapses to L and denote this $K \searrow L$. Note that if $K \searrow L$, then $|L|$ is a strong deformation retract of $|K|$.

Suppose P is a compact *polyhedron* in a p.l. n-manifold M (i.e. P is the image of a finite subcomplex of some allowable triangulation of M). By a *regular neighborhood* of P in M we mean a p.l. n-submanifold N of M such that there is a triangulation (T, L) in the p.l. structure on M and finite subcomplexes K, L of T with $K \searrow L$, $h(|K|) = N$, and $h(|L|) = P$. Note that a regular neighborhood of P may or may not be a neighborhood of P in the traditional sense (i.e. P need not be in the topological interior of N).

1.6. THEOREM. *Let* M *be a p.l. manifold,* (T,h) *a triangulation in the p.l. structure on* M, *and* L *a finite subcomplex of* T. *Let* $N(L,T) = \bigcup_{\sigma \in L} st(\sigma,T)$. *Then* $h(|N(L,T)|)$ *is a regular neighborhood of* $h(|L|)$ *provided*

(i) *Each simplex of* T *which has all its vertices in* L *is in* L (*i.e.* T *is full in* L), *and*

(ii) *If* $\sigma \in N(L,T)$ *and* $\sigma \cap |L| = \emptyset$, *then* $lk(\sigma, N(L,T)) \cap L$ *collapses to a vertex.*

1.7. COROLLARY. $h(|N(L'',T'')|)$ *is a regular neighborhood of* $h(|L|)$; *where* T'' *denotes the second barycentric subdivision of* T.

1.8. THEOREM. *Let* M *be a p.l. manifold,* P *a compact polyhedron in* M *and* N_1 *and* N_2 *regular neighborhoods of* P *in* M, *then*:

(i) *There is a p.l. homeomorphism* $h: N_1 \to N_2$,

(ii) *If* $P \subset \text{Int } N_i (i=1,2)$, *we can require that* $h|P = 1$.

(iii) *If* $N_i \cap \partial M$ *is a regular neighborhood of* $P \cap \partial M (i=1,2)$ (*hence* $N_i \cap \partial M = \emptyset$ *if* $P \cap \partial M = \emptyset$), *there is a p.l. isotopy* $f: M \times I \to M$ *such that* $f_0 = 1$ *and* $f_1(N_1) = N_2$.

(iv) *If, in* (iii), $P \cap M - N_i = \emptyset \ (i=1,2)$, *then we can require that* $f_t|P = 1$ *for all* $t \in I$.

1.9. COROLLARY. *If* (T,h) *is an allowable triangulation of a p.l.* n*-manifold* M *and* L *is a subcomplex of* T *which collapses to a vertex, then any regular neighborhood of* $h(|L|)$ *in* M *is a p.l.* n*-cell.*

1.10. COROLLARY. *If* M *is a p.l.* n*-manifold, then any regular neighborhood of* ∂M *in* M *is p.l. homeomorphic to* $\partial M \times I$.

General Position

A fundamental fact of p.l. topology is that two polyhedra in a p.l. manifold may be moved slightly to be in "general position" in the sense that their intersection is as simple as possible or, more generally, that a map of a polyhedron into a manifold can be approximated by one with simple singularities (self intersections). If a map $f:|K| \to R^n$ embeds the 0-skeleton of K onto a maximally independent set of points and is affine on each simplex of K, then it is a matter of elementary linear algebra to analyze the singularities of f, and this situation serves as a model for general position. However, if we subdivide K in order to make f simplicial (as a map into some triangulation of R^n), it no longer satisfies the above condition – simplexes must be introduced where intersections occur. Since subdivision is a necessity in piecing together local "general position" approximations to yield a global one for a map into a manifold, one is faced with the problem of providing an invariant definition of general position which preserves as much as possible the properties of the above mentioned model. Most of the treatments of general position known to me resolve this problem by accepting a weak definition – usually involving only the dimension of the singular set and neglecting "transversality" of intersections and/or the behavior of the map at a "branch point." Our approach, admittedly cumbersome and inelegant, is to spell out the properties of general position which we will subsequently use. We limit the generality to that actually needed. For a simplicial complex K, a map $f:|K| \to R^n$ is called *affine* if f maps each simplex of K linearly, in terms of barycentric coordinates, into R^n. For a map $f: X \to Y$ we define the *singular set*, $S(f)$, of f to be the closure of $\{x \in X : \#(f^{-1}(f(x))) > 1\}$. We decompose $S(f)$ as a disjoint union, $S(f) = \bigcup_{i \geq 1} S_i(f)$, by $S_i(f) = \{x \in S(f) : \#(f^{-1}(f(x))) = i\}$. Putting $\Sigma_i(f) = f(S_i(f))$, we call the points of $\Sigma_1(f)$ *branch points*, $\Sigma_2(f)$ *double points*, $\Sigma_3(f)$ *triple points*, and so on. For $x \in |K|$, K a simplicial complex, we define the *local dimension* of K at x, locdim(K,x), to be the maximal dimension of the (closed) simplexes of K containing x. A point $x \in |K|$ is called a *regular point* of K, if there is an open neighborhood of x in $|K|$ homeomorphic to either R^q or to R^q_+, $q = \text{locdim}(K,x)$. Regular points of the second type will be called *boundary points* of K.

1.11. DEFINITION. For $k < n \leq 3$, and K a finite k-complex, a map $f: |K| \to R^n$ is in *general position with respect to* K provided:

(i) f is an affine embedding on each simplex of K,

(ii) $\dim S_1(f) \leq n-3$, K has local dimension $(n-1)$ at each point of $S_1(f)$, and $f|(|K|-S_1(f))$ is an immersion.

(iii) for $i \geq 2$, $\dim S_i(f) \leq ik-(i-1)n$ (hence $S_i(f) = \emptyset$ for $i > n$), furthermore for $y \in \Sigma_i(f)$, and $f^{-1}(y) = \{x_1, \cdots, x_i\}$ $\sum_{i=1}^{i} \operatorname{locdim}(K, x_i) \geq (i-1)n$.

(iv) for $i \geq 2$ $S_i(f)$ contains a nonregular point only in the case $n = 3$, $i = 2$. In this case $S_2(f)$ contains only finitely many nonregular points and for each such point x the other point, x_1, of $f^{-1}(f(x))$ is a regular, nonboundary point and K has local dimension 2 at x and at x_1.

(v) for $i \geq 2$ and $y \in \Sigma_i(f)$, $f^{-1}(y)$ contains at most one boundary point; this occurs only when $n = 3$ and K has local dimension 2 at each point of $f^{-1}(y)$.

(vi) for $i \geq 2$ and $y \in \Sigma_i(f)$ a point such that K is regular at each point x_j of $f^{-1}(y)$, f is *transverse at* y in the sense that there exist maximally independent hyperplanes $H_1, \cdots, H_i$ through 0 with $\dim H_j = \operatorname{locdim}(K, x_j)$, a neighborhood N of y in R^n and a p.l. embedding $h: N \to R^n$ with $h(y) = 0$, and with hf taking a neighborhood of x_j in K onto a neighborhood of 0 in H_j or H_j^+ according as x_j is not or is a boundary point of K.

1.12 LEMMA. *Suppose* K *is a finite complex of dimension* $k < n \leq 3$, A, B, C *are subcomplexes of* K *with* $K = A \cup B \cup C$, $A \cup B$ *a full subcomplex of* K *and* $A \cap C = \emptyset$. *Then given any affine map* $g: |K| \to R^n$ *such that* $g||B|$ *is in general position with respect to* B *and given* $\varepsilon > 0$ *there exists an affine map* $f: |K| \to R^n$ *satisfying*:

(a) $d(f(x), g(x)) < \varepsilon$ *for all* $x \in |K|$,

(b) $f||A \cup B| = g||A \cup B|$,

(c) $f||B \cup C|$ *is in general position with respect to* $B \cup C$, *and*

(d) *for each subcomplex* L *of* K *such that* $g||L|$ *is an embedding*, $f||L|$ *is also an embedding.*

PROOF. Choose an ordering $v_1, v_2, \cdots, v_{r+s}$ of the vertices of K with $v_1, \cdots, v_r$ the vertices of $A \cup B$. Let K_q be the subcomplex of K consisting of all simplexes of K whose vertices are in $\{v_1, \cdots, v_{r+q}\}$. Note that $K_0 = A \cup B$ is full in K by assumption and that $K_q (q \geq 1)$ is full in K by construction. Choose $\delta > 0$ such that $\delta < \varepsilon/6$ and such that if σ, τ are simplexes of K with $g(\sigma) \cap g(\tau) = \emptyset$, then $\delta < \frac{1}{6} d(g(\sigma), g(\tau))$. Let $f_0 = g | |K_0| : |K_0| \to R^n$. Proceeding inductively, having defined $f_q : |K_q| \to R^n (q \geq 0)$, let T_q be a complex in R^n with $|T_q| = f_q(|K_q|)$ and let w_{q+1} be a point such that $d(w_{q+1}, g(v_{q+1})) < \delta$, such that w_{q+1} does not lie in any hyperplane of dimension less than n spanned by any set of vertices of T_q, and, in case $n = 3$, w_{q+1} does not lie on any line which meets the interior of each of three distinct 1-simplexes of T_q. Put $f_{q+1} | |K_q| = f_q$, $f_{q+1}(v_{q+1}) = w_{q+1}$, and extend linearly on each simplex of $K_{q+1} - K_q$ to obtain $f_{q+1} : |K_{q+1}| \to R^n$.

We claim that $f = f_s$ in the desired map. Conclusions (a) and (b) follow by construction and choice of δ. For (d) we note that for any simplex σ of K either $f|\sigma = g|\sigma$ or $f|\sigma$ is an embedding. Furthermore, choice of δ yields that $f(\sigma_1) \cap f(\sigma_2) = \emptyset$ whenever $g(\sigma_1) \cap g(\sigma_2) = \emptyset$. So suppose L is a subcomplex of K with $g| |L|$ an embedding. If $f| |L|$ is not an embedding, then there exist simplexes σ_1, σ_2 of L with $f(\mathrm{Int}\, \sigma_1) \cap f(\mathrm{Int}\, \sigma_2) \neq \emptyset$. Choose such a pair with $\dim \sigma_1 + \dim \sigma_2$ minimal. By the above comments $\tau = \sigma_1 \cap \sigma_2$ is a nonempty proper face of each. Let τ_i be the face of σ_i opposite τ. Now $\tau_1 \cap \tau_2 = \emptyset$; so $g(\tau_1) \cap g(\tau_2) = \emptyset$; so $f(\tau_1) \cap f(\tau_2) = \emptyset$. Convexity arguments show that one of $f(\mathrm{Int}\, \tau_1) \cap f(\mathrm{Int}\, \sigma_2)$, $f(\mathrm{Int}\, \sigma_1) \cap f(\mathrm{Int}\, \tau_2)$ must be nonempty contradicting the choice of σ_1, σ_2 and completing the proof of (d).

For (c) we prove inductively that $f_q| |D_q|$ is in general position with respect to $D_q = B \cup (C \cap K_q)$. This is hypothesis for $q = 0$. Assuming $f_q| |D_q|$ to be in general position we consider the conditions of Definition 1.11 for $f_{q+1} | |D_{q+1}|$. Condition (i) follows directly. For (ii) we note that if v is a branch point of f_{q+1}, then v is a vertex of K_{q+1}

and $\operatorname{locdim}(K_{q+1}, v) = 2$, hence $n = 3$. If v is any vertex of K_{q+1} and $f_{q+1}|\operatorname{St}(v, K_{q+1})$ is not an embedding, then v is a "local branch point" of f_{q+1}, i.e. $f_{q+1}^{-1}(f_{q+1}(v)) \cap \operatorname{St}(v, K_{q+1}) = v$. If v is not a branch point of f_{q+1}, then there is a simplex σ of K_{q+1} not containing v with $f_{q+1}(v) \in f_{q+1}(\sigma)$. By induction either $v = v_{q+1}$ or v_{q+1} is a vertex of σ. Either possibility contradicts the choice of w_{q+1}. Thus, if v is not a branch point of f_{q+1}, $f_{q+1}|\operatorname{St}(v, K_{q+1})$ is an embedding. (iii) follows from the observations:

$$\Sigma_2(f_{q+1}) \subset \Sigma_2(f_q) \cup (|T_q| \cap f_{q+1}(\operatorname{St}(v_{q+1}, K_{q+1}))) \cup \Sigma_2(f_{q+1}|\operatorname{St}(v_{q+1}, K_{q+1}))$$

$$\Sigma_i(f_{q+1}) \subset \Sigma_i(f_q) \cup (\Sigma_{i-1}(f_q) \cap f_{q+1}(\operatorname{St}(v_{q+1}, K_{q+1}))) \cup \Sigma_i(f_{q+1}|\operatorname{St}(v_{q+1}, K_{q+1}))$$

for $i > 2$, together with the choice of w_{q+1}. In analyzing $\Sigma_i(f_{q+1}|\operatorname{St}(v_{q+1}, K_{q+1}))$, it is necessary to consider conditions (iv)-(vi) applied to f_q in order to examine $f_{q+1}|\operatorname{lk}(v_{q+1}, K_{q+1})$. Conditions (iv), (v), and (vi) all follow from the choice of w_{k+1}, and the fact that f_q maps each of: the set of nonregular points of K_q, the set of boundary points of K_q and the set of singular points of K_q to a subcomplex of T_q.

1.13 DEFINITION. For K a finite complex of dimension $k < n \leq 3$ and M a p.l. n-manifold, a map $f : |K| \to \operatorname{Int} M$ is in *general position with respect to* K provided there is a finite collection $B_1, \cdots, B_p$ of p.l. n-cells in M with $f(|K|) \subset \cup \operatorname{Int} B_i$ and affine embeddings $h_i : B_i \to R^n$ with $f^{-1}(B_i)$ a subcomplex of K and $h_i f | f^{-1}(B_i) : f^{-1}(B_i) \to R^n$ a general position map in the sense of 1.11 for $i = 1, \cdots, p$.

We may now use Lemma 1.12 to piece together local general position approximations. We note that conditions (b) and (c) are necessary for this process but can, at no expense, be carried along as conclusions; giving

1.14 THEOREM. *Suppose* K *is a finite complex of dimension* $k < n \leq 3$, A, B, C *are subcomplexes of* K *with* $K = A \cup B \cup C$, *and* $A \cap C = \emptyset$.

Then given an n*-manifold* M, *a* p.l. *map* $g : |K| \to M$ *with* $g| \, |B|$ *in general position with respect to* B ($g(|B|) \subset \text{Int } M$), *and* $\varepsilon > 0$ *there exists a* p.l. *map* $f : |K| \to M$ *satisfying*

(a) $d(f(x), g(x)) < \varepsilon$ *for all* $x \in |K|$,

(b) $f| \, |A \cup B| = g|A \cup B|$,

(c) $f| \, |B \cup C|$ *is in general position with respect to some subdivision of* $B \cup C$, *and*

(d) *for each subcomplex* L *of* K *such that* $g| \, |L|$ *is an embedding;* $f| \, |L|$ *is an embedding.*

We conclude by noting that other versions of general position approximation theorems can be deduced from 1.14. We give two examples.

First, to approximate a map $g : (F, \partial F) \to (M, \partial M)$ of manifold pairs one may, using 1.14 applied to $g|\partial F$ together with product structures on neighborhoods of the respective boundaries, obtain an approximation g_1 which is in "general position" when restricted to a neighborhood of ∂F. One may then judiciously apply 1.14 to the double of g_1 which maps the double of F along ∂F to the double of M along ∂M to obtain a further approximation which restricts to a "general position" map $f : (F, \partial F) \to (M, \partial M)$.

As a second example we note that it is frequently desirable to approximate a map $g : |K| \to M$ which is simultaneously in general position in the sense of 1.13 and in "general position" with respect to some polyhedron $P \subset M$. This may be done by approximating the map g_1 of the disjoint union $|K| \cup P$ given by $g_1| \, |K| = g$, $g_1|P = i$ and noting that the approximation may be fixed on P.

CHAPTER 2
HEEGAARD SPLITTINGS

There are several ways of decomposing a 3-manifold as a union of "simple" pieces. The point of this is to analyze the manifold in terms of the manner in which the pieces are attached and thus to reduce the study of the manifold to (more subtle) problems about 2-manifolds. The subject of this chapter is the earliest of these methods to receive attention. It has a simple description and has visual geometric appeal. However, it has proved to be less useful than other methods, e.g. connected sums, and (most particularly) hierarchies which we will consider in subsequent chapters.

A compact, not necessarily connected $(n-1)$-manifold F is *2-sided in* M if there is an embedding $h: F \times [-1,1] \to M$ with $h(x,0) = x$ for all $x \in F$ and $h(F \times [-1,1]) \cap \partial M = h(\partial F \times [-1,1])$.

2.1 LEMMA. *If* F *is a compact* $(n-1)$*-manifold properly embedded in an* n*-manifold* M *and if image* $(i_*: H_1(F; Z_2) \to H_1(M; Z_2)) = 0$, *then* F *is* 2*-sided in* M.

PROOF. By regular neighborhood theory, specifically Corollary 1.10, it suffices to show that each component C of F separates some connected neighborhood of C. If this is not the case then there is a 1-sphere $J \subset M$ such that $J \cap F$ is a single point, with transverse intersection. We may choose J close enough to F so that J is homologous to zero (mod 2) in M. This contradicts the homological invariance (mod 2) of intersection numbers.

Two-sided:

For F a 2-sided $(n-1)$-manifold in an n-manifold M and $h: F\times[-1,1] \to M$ an embedding as above, the n-manifold $R = M - h(F\times(-1,1))$ is called *the result of cutting* M *along* F. By uniqueness of regular neighborhoods (1.8) R is well defined. We further note that M is obtained as an identification space of R: specifically, using product neighborhoods of $h(F\times -1)$ and $h(F\times 1)$ in R one obtains a map $g: R \to M$ such that $g|R-h(F\times\{-1,1\})$ is a homeomorphism onto $M - F$, and for $x \in F, g(h(x,-1)) = g(h(x,1)) = x$.

Cubes with Handles

A 3-manifold M, which contains a collection $\{D_1,\cdots,D_n\}$ of pairwise disjoint, properly embedded 2-cells (2-sided by 2.1) such that the result of cutting M along $\cup D_i$ is a 3-cell is called a *cube with* n-*handles*. By Van Kampen's theorem $\pi_1(M)$ is a free group of rank n.

2.2. THEOREM. *Suppose* $M_i (i=1,2)$ *is a cube with* n_i-*handles. Then* M_1 *is homeomorphic to* M_2 *if and only if* $n_1 = n_2$ *and either both* M_1 *and* M_2 *are orientable or both are nonorientable.*

PROOF. The necessity is clear. For sufficiency let $n_1 = n_2 = n$ and let $h_i:\left(\bigcup_{j=1}^{n} D_{ij}\right)\times[-1,1] \to M_i$ describe M_i as a cube with handles; so $R_i = M_i - h_i(\cup D_{ij}\times(-1,1))$ is a 3-cell. Now M_i will be orientable if and only if for each j the identification of $h_i(D_{ij}\times -1)$ with $h_i(D_{ij}\times 1)$ reverses orientation in terms of the orientations induced by an orientation for R_i. So if both M_1 and M_2 are orientable, there is by 1.5 a homeomorphism $f: R_1 \to R_2$ such that $f(h_1(D_{1j}\times \pm 1) = h_2(D_{2j}\times \pm 1)$. Using 1.4 one can extend f to a homeomorphism of M_1 to M_2 taking $h_1(D_{ij}\times[-1,1])$ to $h_2(D_{2j}\times[-1,1])$.

The case in which M_1 and M_2 are nonorientable is somewhat more involved and might even appear to be false since one might think that M_1 and M_2 could have different numbers of nonorientable handles. Indeed one cannot in general find a homeomorphism of M_1 to M_2 taking D_{1j}

to D_{2j}. However, by changing the collection $\{D_{i1},\cdots,D_{in}\}$ we see that one nonorientable handle suffices. Specifically suppose that the identification of $h_i(D_{ij}\times -1)$ with $h_i(D_{ij}\times 1)$ preserves orientation for $1 \leq j \leq r_i$ and reverses orientation for $r_i < j$ $(r_i \geq 1)$. Let E_i be a properly embedded 2-cell in R_i which separates $h_i(D_{i1}\times -1) \cup \cdots \cup h_i(D_{ir_i}\times -1)$ from the remaining $h_i(D_{ij}\times \pm 1)$'s.

By 1.3 $\{E_i, D_{i2},\cdots,D_{in}\}$ cuts M_i into a 3-cell with exactly one "nonorientable" handle to be attached – the one corresponding to E_i. Following the proof in the orientable case, there is a homeomorphism of M_1 to M_2 taking E_1 to E_2 and D_{1j} to D_{2j} $j \geq 2$.

2.3. THEOREM. *If* F *is a compact, connected 2-manifold with* $\partial F \neq \emptyset$, *then* $F \times I$ *is a cube with* n-*handles*, $n = 1 - \chi(F)$, *which is orientable if and only if* F *is orientable.*

PROOF. There exist $n = 1-\chi(F)$ pairwise disjoint properly embedded 1-cells $\{A_1,\cdots,A_n\}$ in F which cut F to a 2-cell. The 2-cells $\{A_1 \times I,\cdots,A_n \times I\}$ cut $F \times I$ to a 3-cell.

Note 2.3 gives simple examples of nonuniqueness of factors in a cartesian product. For example, if F_1 is a sphere with 3 holes and F_2 is a torus with one hole, then $F_1 \times I = F_2 \times I$ (both an orientable cube with 2-handles).

2.4. THEOREM. *If* Γ *is a connected, finite* 1-*complex in a* 3-*manifold* M, *then any regular neighborhood of* Γ *in* M *is a cube with* n-*handles*, $n = 1-\chi(F)$.

PROOF. Suppose Γ is a subcomplex of a triangulation K of M. By uniqueness of regular neighborhoods, it suffices to consider the regular neighborhood $N = N(\Gamma, K'')$ (cf. 1.7) and to assume $\Gamma \subset \text{Int}\, M$. Let T be a maximal tree in Γ, so there are exactly $n = 1-\chi(\Gamma)$ 1-simplexes $\sigma_1,\cdots,\sigma_n$ of Γ not in T. Now $C = N(T, K'')$ is a 3-cell by 1.9. Let

b_i be the barycenter of σ_i; then $B_i = St(b_i, K'')$ is a 3-cell. Furthermore, $B_i \cap C = \partial B_i \cap \partial C$ is a disjoint union of two 2-cells $D_{i,-1}$, $D_{i,1}$ where $D_{i,j}$ is the star in ∂C of a vertex of σ_i''. Thus there is a homeomorphism $h_i : B^2 \times [-1,1] \to B_i$ with $h(B^2 \times -1) = D_{i,-1}$ and $(B^2 \times 1) = D_{i,1}$. The collection $\{h_i(B^2 \times 0)\}$ cuts N into the 3-cell C.

Splittings and Diagrams

A *Heegaard Splitting* of a closed, connected 3-manifold M is a pair (V_1, V_2) where V_i is a cube with handles $(i = 1,2)$, $M = V_1 \cup V_2$, and $V_1 \cap V_2 = \partial V_1 = \partial V_2$.

We note that the boundary of a cube with n-handles, V, is a closed surface of Euler characteristic $2-2n$ which is orientable if and only if V is orientable. Thus for a Heegaard Splitting (V_1, V_2) of a 3-manifold M, V_1 and V_2 have the same number (called the *genus of the splitting*) of handles and either both are orientable or both are nonorientable according as M is orientable or nonorientable.

2.5. THEOREM. *Each closed, connected* 3-*manifold* M *has a Heegaard Splitting.*

PROOF. Let K triangulate M. Let Γ_1 be the 1-skeleton of K and Γ_2 be the dual 1-skeleton, i.e. Γ_2 is the maximal 1-subcomplex of K' disjoint from Γ_1. Let $V_i = N(\Gamma_i, K'')$. By 1.7 V_1 is a regular neighborhood of Γ_1 and thus by 2.4 a cube with handles. We must appeal to 1.6 to show that V_2 is a regular neighborhood of Γ_2. One verifies the conditions $M = V_1 \cup V_2$, $V_1 \cap V_2 = \partial V_1 = \partial V_2$ to complete the proof.

Suppose we have a Heegaard Splitting (V_1, V_2) of a 3-manifold M. Let $\{D_1, \cdots, D_n\}$ be any collection of pairwise disjoint properly embedded 2-cells in V_2 which cut V_2 into a 3-cell. The pairwise disjoint 1-spheres $\{\partial D_1, \cdots, \partial D_n\}$ cut $\partial V_2 = \partial V_1$ into a 2-sphere with 2n holes. We call the system $(V_1; \partial D_1, \cdots, \partial D_n)$ a *Heegaard diagram associated with the Heegaard Splitting* (V_1, V_2). We can recover M from a Heegaard diagram; more generally, given a cube with n-handles, V_1, and a collection $\{J_1, \cdots, J_n\}$ of pairwise disjoint, 2-sided 1-spheres in ∂V_1 which cut ∂V_1 into a 2-sphere with 2n holes, then $(V_1; J_1, \cdots, J_n)$ is a Heegaard diagram associated with a Heegaard Splitting of a uniquely determined closed 3-manifold M. One constructs M as follows. For each $i = 1, \cdots, n$ attach a copy of $B^2 \times I$ to V_1 by identifying $\partial B^2 \times I$ with a neighborhood of J_i in ∂V_1. The resulting manifold, M_1, has a 2-sphere boundary. We obtain M by attaching a copy of B^3 to M_1 by identifying ∂B^3 with ∂M_1. By construction $V_2 = \overline{M - V_1}$ is a cube with handles and by the isotopy theorems (1.4, 1.5) M is unique up to homeomorphism.

Given a Heegaard diagram $(V_1; J_1, \cdots, J_n)$ as above we can construct a *presentation* for $\pi_1(M)$ as follows. Choose a free basis $\{x_1, \cdots, x_n\}$ for the free group $\pi_1(V_1)$. For $i = 1, \cdots, n$ let r_i be a word in $x_1, \cdots, x_n$ representing an element of $\pi_1(V_1)$ determined by J_i (r_i is unique up to inversion and conjugation). By Van Kampen's theorem $< x_1, \cdots, x_n : r_1, \cdots, r_n >$ is a presentation for $\pi_1(M_1) \cong \pi_1(M)$.

While Heegaard diagrams and splittings provide a convenient way of viewing, and constructing, 3-manifolds, they have limited use in the systematic study of 3-manifolds. The main reason for this lies in the fact that there are many different Heegaard diagrams associated with a given Heegaard splitting (of genus ≥ 2). We illustrate this in

2.6. EXERCISE. *Let* $(V_1; J_1, J_2)$ *be the Heegaard diagram shown in Fig. 1 and* M *the associated* 3-*manifold. Observe* $\pi_1(M) = < x_1, x_2 : x_1^2 x_2^3, x_1^3 x_2^4 > = 1$. *Show that* $M = S^3$ *by verifying that* $(V_1; J_1, J_2)$ *and* (V'_1, J'_1, J'_2) *(Fig. 2) determine the same Heegaard splitting.*

2.7. EXERCISE. *Let* $(V_1; J_1, J_2)$ *be the Heegaard diagram shown in Fig. 3 and* M *the associated* 3-*manifold. Note that*

$$\pi_1(M) = < x_1, x_2 : x_1 x_2 x_1 x_2^{-1} x_1^{-1} x_2^{-1},\ x_1 x_2 x_1^{-2} x_2 x_1 x_2^{-1} > .$$

Verify that the commutator quotient group of $\pi_1(M)$ *is trivial; hence that* $H_1(M) = 0$ *and thus* $H_2(M) = 0$, $H_3(M) = Z$ *(by Poincaré duality). Thus*

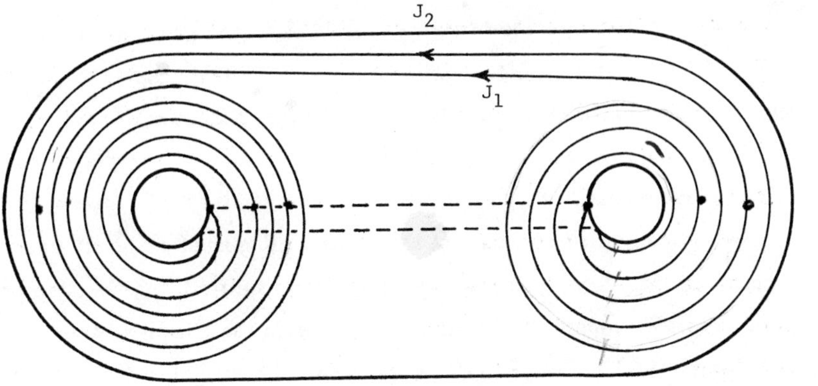

Fig. 1

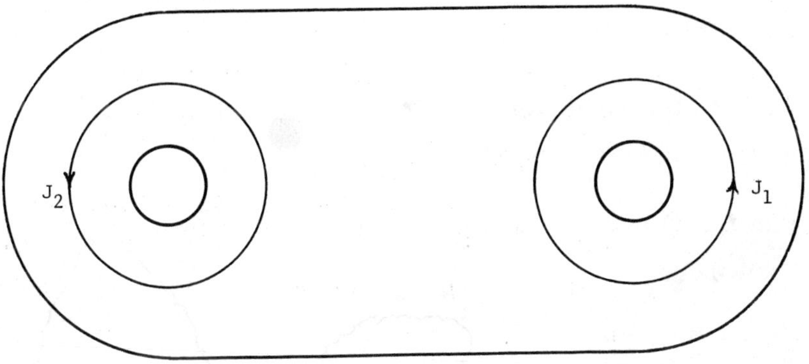

Fig. 2

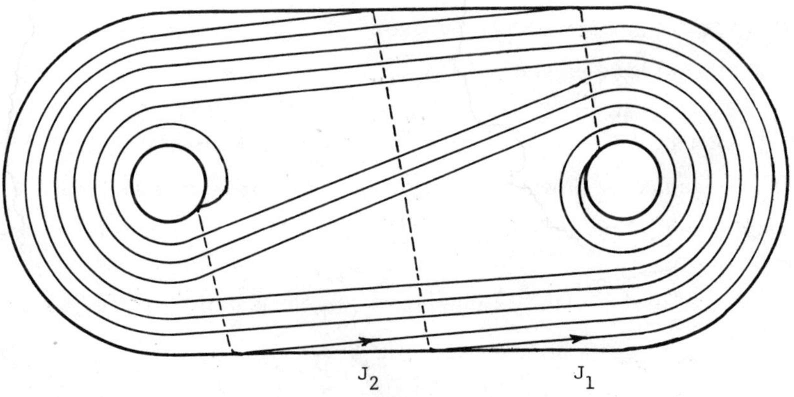

Fig. 3

M *cannot be distinguished from* S^3 *by homology; such a manifold will be called a homology 3-sphere. Show that* $x_1 \to (12345)$, $x_2 \to (15324)$ *determines a nontrivial representation of* $\pi_1(M)$ *and hence* $\pi_1(M) \neq 1$.

Genus One Splittings

Genus one splittings can be effectively classified. Clearly only S^3 admits a genus zero splitting. For genus one we first consider the orientable case. We regard S^1 as the unit circle in the complex plane with base point 1. The loops $\alpha, \beta : (I, \partial I) \to (S^1 \times S^1, x_0)$, $x_0 = (1,1)$ given by $\alpha(t) = (e^{2\pi i t}, 1)$, $\beta(t) = (1, e^{2\pi i t})$ determine a free basis $a = [\alpha]$, $b = [\beta]$ for the free abelian group $\pi_1(S^1 \times S^1, x_0)$. An oriented loop in $S^1 \times S^1$ determines a unique element (regardless of how it is joined to the base point) of the abelian group $\pi_1(S^1 \times S^1)$.

2.8. LEMMA. *An element* $1 \neq a^p b^q \in \pi_1(S^1 \times S^1)$ *is represented by a simple loop if and only if* $(p,q) = 1$.

PROOF. If $(p,q) = 1$, define $\omega(t) = (e^{2p\pi i t}, e^{2q\pi i t})$. Then ω is simple and $[\omega] = a^p b^q$.

Given a simple loop $\omega : (I, \partial I) \to (S^1 \times S^1, x_0)$, $[\omega] \neq 1$, the 1-sphere $\omega(I)$ cuts $S^1 \times S^1$ into an annulus. Thus there is a homeomorphism $h : S^1 \times S^1 \to S^1 \times S^1$ such that $h \circ \alpha = \omega$. Now $h_*(a) = [\omega] = a^p b^q$ and $h_*(b) = a^r b^s$ for some p, q, r, s. Since h_* is an automorphism of the free abelian group $\pi_1(S^1 \times S^1)$, $\det \begin{pmatrix} p & q \\ r & s \end{pmatrix} = \pm 1$. This completes the proof. Actually we have proved

2.9. LEMMA. *A pair* $a^p b^q$, $a^r b^s$ *of nontrivial elements of* $\pi_1(S^1 \times S^1)$ *is represented by a pair of simple loops which meet transversely in a single point if and only if* $\det \begin{pmatrix} p & q \\ r & s \end{pmatrix} = \pm 1$.

The following result is classical. An elementary proof can be found in [21].

2.10. LEMMA. *If* F *is any 2-manifold and* $f_0, f_1 : S^1 \to F$ *are homotopic embeddings, which are not null homotopic, then there is an isotopy* $g : F \times I \to F$ *such that* $g_0 = 1$ *and* $g_1 \circ f_0 = f_1$.

Let $V_1 = S^1 \times B^2$. We have generators a, b for $\pi_1(\partial V_1) = \pi_1(S^1 \times S^1)$ as above with $\ker(\pi_1(\partial V_1) \to \pi_1(V_1))$ the subgroup generated by b. Given $(p, q) = 1$, $p \geq 0$, there is, by the above, a unique 3-manifold $L_{p,q}$ determined by the Heegaard diagram $(V_1; J)$ where J is (any) oriented 1-sphere in ∂V_1 representing the element $a^p b^q$ in $\pi_1(\partial V_1)$. $L_{p,q}$ is called the *lens space of type* p,q, although $S^3 = L_{1,0}$ and $S^2 \times S^1 = L_{0,1}$ are not usually considered as lens spaces. We note that $\pi_1(L_{p,q}) \cong Z_p$, where we define $Z_1 = 1$ and $Z_0 = Z$.

We note that, by 2.10, a homeomorphism $h : \partial V_1 \to \partial V_1$ extends to a homeomorphism of V_1' to itself if and only if $h_*(b) = b^{\pm 1}$. Using this fact, a little calculation will establish

2.11. EXERCISE.

(i) $L_{1,q} = S^3$ *for all* q.

(ii) $L_{p,q} = L_{p,q'}$ *provided either*

(a) $q \equiv \pm q' \bmod p$, *or*

(b) $qq' \equiv \pm 1 \bmod p$.

One actually shows that in (ii) a homeomorphism $h : L_{p,q} \to L_{p,q'}$ can be constructed which preserves the "halves" of the Heegaard Splitting in case (a) and reverses them in case (b). Indeed conditions (a) or (b) are necessary for such a homeomorphism. Of course an arbitrary homeomorphism between manifolds need not preserve Heegaard Splittings in any sense; however, it is true that (a), (b) are also necessary conditions in general. A proof of this fact requires additional machinery.

We turn to the nonorientable case. First we review some facts about nonorientable manifolds.

Given a n-manifold M and a path $\alpha : I \to M$ choose a subdivision $0 = t_0 < t_1 < \cdots < t_k = 1$ of I such that $f([t_{i-1}, t_i])$ lies in a Euclidean neighborhood N_i in M. By translating in N_1 a neighborhood of $\alpha(t_0)$ in N_1 to a neighborhood of $\alpha(t_1)$ in $N_1 \cap N_2$ and so on we obtain a homeomorphism from a neighborhood U of $\alpha(0)$ to a neighborhood V of $\alpha(1)$. If α is a loop, we say α is *orientation preserving* if (some such) h is an orientation-preserving homeomorphism. It is a straightforward matter to check that this notion is well defined and independent of the homotopy class of α. Furthermore the product of two loops is orientation reversing if and only if exactly one of them is orientation reversing. We put $\omega(M) = \{[\alpha] \in \pi_1(M) : \alpha$ is orientation preserving$\}$. If M is orientable, then $\omega(M) = \pi_1(M)$; otherwise we have

2.12. THEOREM. *Let* M *be a nonorientable manifold. Then* $\omega(M)$ *is a subgroup of index two in* $\pi_1(M)$ *(hence normal). If* $p : \tilde{M} \to M$ *is the (regular) covering space with* $p_*(\pi_1(\tilde{M})) = \omega(M)$, *then* $\tilde{M}$ *is orientable and minimal in the sense that for any other covering space* $p' : M' \to M$ *with* M' *orientable, there is a covering projection* $q : M' \to \tilde{M}$ *with* $p' = p \circ q$. $\tilde{M}$ *is called the orientable double cover of* M.

Now let K be the Klein bottle as pictured below:

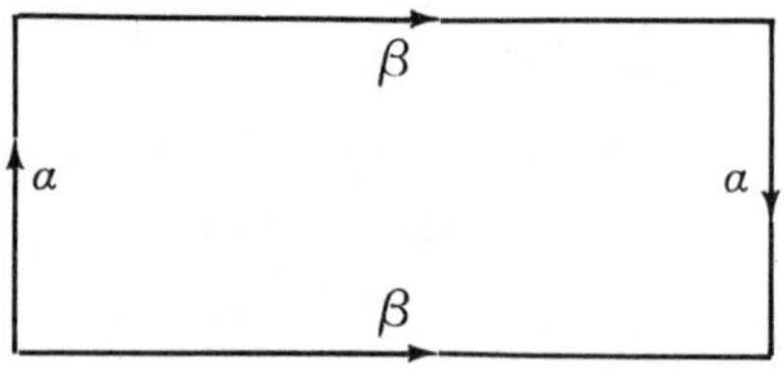

Then α, β are simple loops which intersect transversely in a single point. We have generators $a = [\alpha]$, $b = [\beta]$ for $\pi_1(K) = \langle a, b : bab^{-1} = a^{-1} \rangle$. Furthermore α is orientation preserving, β is orientation reversing, $\omega(K)$ is a free abelian of rank 2 generated by a and b^2 and the orientable double cover is $S^1 \times S^1$. The subgroup generated by a is normal with the subgroup generated by b as a complement i.e. $\pi_1(M)$ is a semi-direct product of Z by Z. Thus every element of $\pi_1(K)$ has a unique expression in the form $a^m b^n$ which can be computed using the relations

$$b^r a^s = \begin{cases} a^s b^r & r \text{ even} \\ a^{-s} b^r & r \text{ odd} \end{cases}$$

Now $\pi_1(K)$ is nonabelian, so an oriented loop determines an element of $\pi_1(K)$ unique up to conjugacy. However, two oriented loops are freely homotopic if and only if they determine the same conjugacy class. By 2.10 two oriented (nontrivial) simple loops are isotopic if and only if they determine the same conjugacy class. Corresponding to 2.9 we have

2.13. LEMMA. *A pair* $a^p b^q$, $a^r b^s$ *of nontrivial elements of* $\pi_1(K)$ *is represented up to conjugacy by a pair* ϕ, ψ *of simple loops which meet transversely in a single point with* ϕ *orientation preserving and* ψ *orientation reversing if and only if* $p = \pm 1$, $q = 0$, *and* $s = \pm 1$.

PROOF. Sufficiency is clear. For necessity, suppose ϕ, ψ are given. Then a neighborhood N of $\psi(I)$ is a Möbius band and $\overline{K - N}$ is also a Möbius band. $\phi(I)$ cuts each of these into a rectangle. Thus there is a homeomorphism $h : K \to K$ such that $h \circ \alpha = \phi$, $h \circ \beta = \psi$. Putting $h_*(a) = a^p b^q$ and noting that ϕ, ψ represent the conjugacy classes of $h_*(a)$ and $h_*(b)$ respectively, we see that q is even and s is odd. Now h_* must preserve the relation $bab^{-1} = a^{-1}$. Since $h_*(bab^{-1}) = a^r b^s a^p b^q b^{-s} a^{-r} = a^{-p} b^q$, and since $h_*(a^{-1}) = b^{-q} a^{-p} = a^{-p} b^{-q}$ we must have $q = 0$. Under the orientable double covering α and β^2 lift to standard generators for $\pi_1(S' \times S')$. ψ^2 lifts to a simple loop which meets each of the two liftings of ϕ transversely in a single point. By 2.9 $\det \begin{pmatrix} p & 0 \\ r & s \end{pmatrix} = \pm 1$. Hence $p = \pm 1$, $s = \pm 1$.

From this, one easily establishes

2.14. EXERCISE. *There is exactly one nonorientable 3-manifold with a genus one Heegaard Splitting: the nonorientable 2-sphere bundle over* S^1.

CHAPTER 3
CONNECTED SUMS

Suppose M, M_1, M_2 are connected 3-manifolds and that there are 3-cells $B_i \subset \text{Int } M_i (i=1,2)$ and embeddings $h_i : R_i \to M$ $(R_i = M_i - \text{Int } B_i)$ with $h_1(R_1) \cap h_2(R_2) = h_1(\partial B_1) = h_2(\partial B_2)$ and $M = h_1(R_1) \cup h_2(R_2)$. We say that M is *a connected sum of* M_1 and M_2 and denote this $M = M_1 \# M_2$. If we try to reconstruct M from M_1 and M_2, we see by 1.5 that we may choose the 3-cell $B_i \subset \text{Int } M_i$ at will and by 1.4 that there are only two essentially different ways of identifying ∂B_1 with ∂B_2. If M, M_1, M_2 are oriented we require that $h_i : R_i \to M$ be orientation preserving (equivalently that the identification of ∂B_1 with ∂B_2 reverse the induced orientations). With this restriction we have

3.1. LEMMA. *Connected sum is a well-defined associative and commutative operation in the category of oriented* 3-*manifolds and orientation preserving homeomorphisms.*

On the other hand we will show, at the end of this chapter, that there exist oriented 3-manifolds M_1, M_2 with $M_1 \# M_2$ not homeomorphic to $M_1 \# (-M_2)$ ($-M_2$ denotes M_2 with the opposite orientation). We will continue to use the terminology and notation of connected sums in the unoriented category and will emphasize the possible ambiguity by the term *a connected sum.* Observe that there are in any case, at most two homeomorphism types for $M_1 \# M_2$ and only one if one of M_1, M_2 admits a self homeomorphism which fixes some point and reverses orientation on a neighborhood of this point.

It is convenient to use the notation $\hat{M}$ for the manifold obtained from the 3-manifold M by capping off each 2-sphere component of ∂M with a 3-cell. Thus $M \subset \hat{M}$, the closure of each component of $\hat{M} - M$ is a 3-cell, $\partial\hat{M}$ contains no 2-sphere, and the inclusion induced map $i_* : \pi_1(M) \to \pi_1(\hat{M})$ is an isomorphism. The last part follows from Van Kampen's theorem as does:

3.2. LEMMA. *If* $M_1 \# M_2$ *is a connected sum decomposition of a 3-manifold* M, *then there is a natural isomorphism*

$$\pi_1(M) \cong \pi_1(M_1) * \pi_1(M_2)\,.$$

Our goal for the remainder of this chapter is to prove an existence and (properly stated) uniqueness theorem for a prime decomposition (via #) of a compact 3-manifold. This, coupled with a converse to 3.2 (for M closed) to be proved in Chapter 7, establishes an important tool in the study of 3-manifolds which is enhanced by the fact (Chapter 4) that "most" prime 3-manifolds are aspherical ($\pi_i(M) = 0$ for $i \geq 2$).

The existence of a prime decomposition "almost" follows from 3.2; for if $p(G) = \inf\{\#(A);\ A \text{ generates } G\}$, we have

3.3. LEMMA. *If* $G_1 * G_2$ *is a finitely generated group, then* $p(G_1 * G_2) = p(G_1) + p(G_2)$.

This follows directly from

3.4. GRUSHKO'S THEOREM. *If* $G_1 * G_2$ *is a finitely generated group,* F *is a finitely generated free group and* $\eta : F \to G_1 * G_2$ *is an epimorphism, then there exist free groups* $F_i (i=1,2)$, *homomorphisms* $\phi_i : F_i \to G_i$, *and an isomorphism* $\psi : F \to F_1 * F_2$ *such that* $(\phi_1 * \phi_2) \circ \psi = \eta$.

We refer to [63, p. 191] or [66, p. 225] for a proof of 2.4.

By 3.3 we see that a compact 3-manifold M cannot have more than $p(\pi_1(M))$ nonsimply connected factors. However, the possibility that M has an unbounded number of simply-connected factors must be ruled out. A few comments about the Poincaré conjecture are appropriate at this point.

In [83] Poincaré asserted that a manifold with trivial homology is simply connected and therefore a sphere: "... est simplement connexe, c'est-a dire homeómorphe a l'hypersphere" [83, p. 308]. Shortly thereafter he discovered an error in his first assertion and gave an example [84] of a nonsimply connected homology 3-sphere (cf. 2.7). However the following is still unsettled.

3.5. THE POINCARÉ CONJECTURE. *Each closed, connected, simply connected 3-manifold is homeomorphic to* S^3.

Using the term *homotopy* n-*sphere* for an n-manifold homotopy equivalent to S^n, we have

3.6. THEOREM. *A 3-manifold* M *is a homotopy* 3-*sphere if and only if* M *is closed, connected, and simply connected.*

PROOF. Necessity is clear. For sufficiency, suppose $\partial M = \emptyset$ and $\pi_0(M) = \pi_1(M) = 1$. Then $H_1(M) = 0$ and $H_2(M) = 0$ by duality. Let B be a 3-cell in M and $C = \overline{M - B}$. By Van Kampen's theorem $\pi_1(C) = 1$. By a Mayer-Vietoris argument $H_2(C) = 0$, and thus $H_q(C) = 0$ for $q \geq 1$. By the Hurewicz theorem $\pi_q(C) = 0$ for all q, and thus C is contractible. Hence a homeomorphism $f_+ : B^3_+ \to B$ extends to a map $f : S^3 \to M$ with $f(B^3_-) = C$. Similarly $f_+^{-1} : B \to B^3_+$ extends to a map $g = M \to S^3$ with $g(C) = B^3_-$. One verifies that f is a homotopy equivalence with with homotopy inverse g.

We use the term *fake* 3-*cell* for a compact, contractible 3-manifold which is not homeomorphic to B^3. By the above argument, the Poincaré conjecture is equivalent to

3.5′. CONJECTURE. *There is no fake 3-cell.*

Primes

A 3-manifold M is *prime* if $M = M_1 \# M_2$ implies one of M_1, M_2 is a 3-sphere. We are interested in prime factorizations of bounded as well as closed 3-manifolds. This causes a mild curiosity when the boundary contains 2-spheres. For example $S^2 \times I = B^3 \# B^3$ is a prime factorization of $S^2 \times I$, whereas $(S^2 \times I)\hat{} = S^3$. More generally we have

3.7. LEMMA. *Suppose* M *is a compact 3-manifold with exactly* k *2-spheres in* ∂M. *If* $\hat{M} = M_1 \# \cdots \# M_q$ *is a prime factorization of* $\hat{M}$, *then* $M = M_1 \# \cdots \# M_q \# B^3 \# \cdots \# B^3$ *is a prime factorization of* M *(where there are* k *factors of* B^3*).*

3.8. LEMMA. *Suppose* M *is a 3-manifold and that* $\operatorname{Int} M$ *contains a 2-sphere* S *with* $M - S$ *connected. Then* $M = M_1 \# M_2$ *where* M_1 *is a 2-sphere bundle over* S^1.

PROOF. There exists a 1-sphere $J \subset \operatorname{Int} M$ such that J meets S transversely in a single point. A regular neighborhood N of $S \cup J$ in $\operatorname{Int} M$ has the form $S \times [-1,1] \cup B$, where B is a 3-cell and $B \cap (S \times [-1,1]) = \partial B \cap \partial(S \times [-1,1])$ consists of a 2-cell in $S \times (-1)$ and a 2-cell in $S \times 1$. Then ∂N is a 2-sphere and N cut along S is homeomorphic to $S^2 \times I$ with the interior of a 3-cell removed. Thus $M_1 = \hat{N}$ is a 2-sphere bundle over S^1 which is a factor of M.

We note that, in the above lemma, M_1 will be orientable (hence $M_1 = S^2 \times S^1$) if and only if J is an orientation preserving simple closed curve. Let P be the nonorientable S^2 bundle over S^1. Using this observation one can establish

3.9. EXERCISE. $P \# P = P \# (S^2 \times S^1)$.

3.10. THEOREM. *If* M *is a compact 3-manifold, then* $M = R\# M_1 \# \cdots \# M_k$ *where each* M_i *is a 2-sphere bundle over* S^1, $0 \le k \le p(\pi_1(M))$, *and each 2-sphere in* Int R *separates* R.

PROOF. Apply 3.2, 3.3 and 3.7.

3.11. LEMMA. *Suppose* M *is a compact 3-manifold,* $M = \hat{M}$, $\pi_1(M)$ *is not a nontrivial free product, and* M *contains no fake 3-cell. Then* M *is prime.*

PROOF. By 3.2 one factor, M_1, in any factorization of M must be simply connected. Now $H_2(M_1, \partial M_1; Z_2) \cong H^1(M_1; Z_2) = 0$. From $0 = H_2(M_1, \partial M_1; Z_2) \to H_1(\partial M_1; Z_2) \to H_1(M_1; Z_2) = 0$, $H_1(\partial M_1; Z_2) = 0$. By assumption ∂M_1 contains no 2-sphere; so M_1 is closed. Hence M_1 is a homotopy 3-sphere. By assumption the part of M_1 in M is not a fake 3-cell; hence $M_1 = S^3$.

Noting that a fake 3-cell in a manifold M lifts to every covering space of M, one can establish

3.12. EXERCISE. *Lens spaces and 2-sphere bundles over* S^1 *are prime.*

A 3-manifold M is *irreducible* if each 2-sphere in M bounds a 3-cell in M. Clearly irreducible 3-manifolds are prime. As a partial converse we have then every sphere separates!

3.13. LEMMA. *If* M *is a prime 3-manifold and* M *is not irreducible, then* M *is a 2-sphere bundle over* S^1.

PROOF. Since M is prime, any separating 2-sphere in Int M bounds a 3-cell in M. Thus Int M must contain a nonseparating 2-sphere and the conclusion follows from 3.8 and 3.12.

Existence of Factorizations

A compact 3-manifold M with $\partial M \neq \emptyset$ and $\hat{M} = S^3$ is called a *punctured* 3-*cell*. The main step in the existence of a prime factorization is due to H. Kneser [57]:

3.14. LEMMA. *Suppose that* M *is a compact* 3-*manifold and that each* 2-*sphere in* Int M *separates* M. *Then there is an integer* k(M) *such that if* $\{S_1, \cdots, S_n\}$ *is any collection of* $n \geq k(M)$ *pairwise disjoint* 2-*spheres in* Int M, *then the closure of some component of* $M - \cup S_i$ *is a punctured* 3-*cell.*

PROOF. We fix a triangulation T of M. If $\{S_1, \cdots, S_n\}$ is a collection of pairwise disjoint 2-spheres in general position with respect to T we define the complexity of $\{S_i\}$ to be the pair (α, β) where $\alpha = \#(T^{(1)} \cap \cup S_i)$ and β is the sum over all 2-simplexes σ of T of the number of components of $\sigma \cap \cup S_i$.

Now suppose that for some integer n we have a collection $\{S_1, \cdots, S_n\}$ of pairwise disjoint 2-spheres in Int M satisfying

(i) the closure of no component of $M - \cup S_i$ is a punctured 3-cell, and

(ii) among all collections of n 2-spheres satisfying (i), $\{S_i\}$ has minimal complexity (under lexicographical ordering).

Let D be a 2-cell in Int M with $D \cap \cup S_i = \partial D$. Then $\partial D \subset S_i$ for some i. Let E′ and E″ be the 2-cells in S_i bounded by ∂D. Put $S_i' = D \cup E'$ and $S''_i = D \cup E''$. We claim that at least one of the collections $\{S_1, \cdots, S_i', \cdots, S_n\}$, $\{S_1, \cdots, S''_i, \cdots, S_n\}$ continues to satisfy (i). This is established using the fact that the union of two punctured 3-cells meeting in a 2-cell in the boundary of each is a punctured 3-cell and the fact that a properly embedded 2-cell in a punctured 3-cell, P, cuts P into two punctured 3-cells. We say that a new collection obtained in the above manner is obtained by a D-*modification* on $\{S_i\}$.

We establish several properties:

(iii) If σ is a 2-simplex of T, then no component of $\cup S_i \cap \sigma$ is a 1-sphere; for if not some such component bounds a 2-cell $D \subset \sigma$ with $\text{Int } D \cap \cup S_i = \emptyset$. By performing a D-modification on $\{S_i\}$ we obtain a new collection which satisfies (i) and which may be adjusted in a neighborhood of D to one of smaller complexity (β is reduced and a is not increased).

(iv) If σ is a 2-simplex of T, then no component of $\cup S_i \cap \sigma$ is a 1-cell with both ends in the same edge of σ; for if not then some such component A together with a subarc B of some edge of σ bounds a 2-cell $D \subset \sigma$ such that $D \cap \cup S_i = A$. Let N be a small regular neighborhood of D such that $N \cap \cup S_i$ is a 2-cell E and such that ∂E bounds a 2-cell E' in ∂N with $E' \cap T^{(1)} = \emptyset$. By replacing E by E' we obtain a new collection with smaller complexity (a has been reduced) which continues to satisfy (i); since it is isotopic to the old one.

(v) If τ is a 3-simplex of T and J is a component of $\partial \tau \cap \cup S_i$, then each component of $\partial \tau - J$ contains a vertex of τ; otherwise we would contradict (iii) or (iv).

(vi) If τ is a 3-simplex of T, then each component of $\tau \cap \cup S_i$ is a 2-cell. If not, then there is a component C of $\cup S_i \cap \tau$ such that C is not a 2-cell but some component J of ∂C bounds a 2-cell E in $\partial \tau$ such that every component of $\tau \cap \cup S_i$ which meets Int E is a 2-cell. One can construct a 2-cell $D \subset \tau$ such that $D \cap \cup S_i = J$ and $\text{Int } D \subset \text{Int } \tau$. By performing a D-modification on $\{S_i\}$ we get a new collection which still satisfies (i). If C is contained in the new 2-sphere, we may modify by an isotopy in a neighborhood of J to reduce a. If C is not contained in the new 2-sphere, then since C has more than one boundary component (each of which meets $T^{(1)}$) we have automatically reduced a.

Now, by (vi), if τ is a 3-simplex of T and X is the closure of a component of $\tau - \cup S_i$, then X is a 3-cell whose boundary is made up of some 2-cell components of $\tau \cap \cup S_i$ together with a connected submanifold of $\partial \tau$. We say that X is *good* if $X \cap \partial \tau$ is an annulus which contains no vertex of τ. We observe that for each 3-simplex τ there are at most 6 bad X's; at most 4 can contain a vertex of τ and, by (v) at most 2 others can fail to meet $\partial\tau$ in an annulus.

Let R be the closure of a component of $M - \cup S_i$; R is *good* if for each 3-simplex τ of T each component of $R \cap \tau$ is good in the above sense. Using (iii)-(vi) we show that if R is good, then R is a fiber bundle over a 2-manifold with fiber I; one builds up a local product structure together with a cross section first over $R \cap T^{(1)}$, then extends over $R \cap T^{(2)}$, and finally over R. Now ∂R contains a 2-sphere. Thus there are just two possibilities for a good $R: S^2 \times I$, or the twisted I-bundle over P^2.

We now put $k(M) = \dim H_1(M; Z_2) + 6t$ where t is the number of 3-simplexes of T. Suppose we have a collection of n 2-spheres satisfying the hypothesis but not the conclusion of our lemma with $n \geq k(M)$. Then some other collection of n 2-spheres satisfies (i) and (ii), hence (iii)-(vi). For this collection $M - \cup S_i$ has $n+1$ components. By the previous remarks, at most 6t of these can be bad. Thus we have at least $\dim H_1(M; Z_2) + 1$ good components. At least one of these is homeomorphic to $S^2 \times I$; giving the final contradiction.

3.15. THEOREM. *Each compact 3-manifold can be expressed as a connected sum of finite number of prime factors.*

PROOF. By 3.10 we can factor $M = R \# M_1 \# \cdots \# M_k$ where each M_i is a 2-sphere bundle over S^1 (hence prime) and each 2-sphere in Int R separates R. We continue to factor R (nontrivially) as long as possible. Note that if we have $R = R_1 \# \cdots \# R_{n+1}$, then there are disjoint 2-spheres $S_1, \cdots, S_n$ in Int R such that if we denote the closures of the components of $R - \cup S_i$, by $Q_1, \cdots, Q_{n+1}$ we have $R_i = \hat{Q}_i$. By 3.14 we see that R cannot have as many as k(R) nontrivial factors. The proof is completed by 3.7.

Uniqueness of Factorizations

In [70], J. Milnor proved a uniqueness theorem for prime decompositions of closed, oriented 3-manifolds. There is no reason to restrict attention to closed manifolds and it is not difficult to see that 3.9 represents the most general nonuniqueness. Thus by properly normalizing, we get a uniqueness theorem in all cases; which we proceed to prove. Keeping in mind the comments at the beginning of this chapter, we continue to work simultaneously in the oriented and unoriented category with the understanding the equality has different meanings in these categories.

3.16. LEMMA. *If* $M = M_1 \# M_2$ *where* M_2 *is a 2-sphere bundle over* S^1, *then there is a nonseparating 2-sphere* $S \subset \text{Int } M$ *such that the manifold obtained by cutting* M *along* S *and capping off the two resulting 2-sphere boundary components with 3-cells is homeomorphic to* M_1.

PROOF. We have 3-cells $B_i \subset \text{Int } M_i$ and embeddings $h_i : M_i - \text{Int } B_i \to M$ as in the definition of #. Let $S = h_2(S^*)$ where S^* is a fiber of M_2 missing B_2. Since S^* cuts M_2 to a product $S^2 \times I$, the lemma follows.

Let P be the nonorientable S^2 bundle over S^1.

3.17. LEMMA. *If* $M = M_1 \# (S^2 \times S^1)$ *where* M_1 *is nonorientable, then* $M = M_1 \# P$ *and conversely.*

PROOF. Assume, for convenience, that $M = \hat{M}$ (i.e. ∂M contains no 2-sphere). Then there is a nonseparating 2-sphere $S \subset \text{Int } M$ and an orientation-preserving simple loop $J \subset \text{Int } M$ which meets S transversely in a single point such that if N is a regular neighborhood of $S \cup J$, then $\hat{N} = S^2 \times S^1$ and $(M - \text{Int } N)\hat{} = M_1$. By multiplying J by an orientation-reversing loop in $M - S$ and adjusting to general position, we obtain an orientation-reversing simple loop K which meets S transversely in a single point. Let R be a regular neighborhood of $S \cup K$ in M. Then we have $M = M'_1 \# P$ where $M'_1 = (M - \text{Int } R)\hat{}$, $P = \hat{R}$. By 3.16 M_1 and

M'_1 are both homeomorphic to $(M - S \times (-1,1))\hat{}$. The converse follows similarily.

3.18. LEMMA. *Let* M *be a 3-manifold and* S_1, S_2 *nonseparating 2-spheres in* Int M. *Suppose that either both* $M - S_1$ *and* $M - S_2$ *are orientable or both are nonorientable. Then there is a homeomorphism* $h: M \to M$ *such that* $h(S_1) = S_2$. *If* M *is oriented we may assume* h *preserves orientation.*

PROOF. We may assume that S_1 is in general position with respect to S_2; so that each component of $S_1 \cap S_2$ is a 1-sphere. We induct on the number k of components of $S_1 \cap S_2$.

If $k = 0$ choose disjoint product neighborhoods $S_1 \times [-1,1]$, $S_2 \times [-1,1]$ of S_1 and S_2 in Int M. If $M - (S_1 \cup S_2)$ is not connected, we choose the notation so that $S_1 \times 1$ and $S_2 \times 1$ lie in the same component of $M - (S_1 \cup S_2)$. Let $R = M - (S_1 \times (-1,1) \cup S_2 \times (-1,1))$. By applying 1.5 to (the components of) $\hat{R}$ and then restricting, we obtain a homeomorphism $g: R \to R$ such that $g(S_1 \times i) = S_2 \times i$, $i = -1, 1$. Using 1.4 we can extend g to the desired homeomorphism h if $g|S_1 \times (-1)$ is homotopic to $g|S_1 \times 1$ in $S_2 \times [-1,1]$. If R is orientable, this condition is forced on us by the assumption that $M - S_1$ and $M - S_2$ be either both orientable or both nonorientable. If R is nonorientable, then g need not extend. However, in this case there is a homeomorphism $f: R \to R$ such that $f|S_1 \times (-1) \cup S_2 \times \{-1,1\} = \mathrm{Id}$ and $f|S_1 \times 1: S_1 \times 1 \to S_1 \times 1$ reverses orientation. Then $f \circ g$ does extend to the desired homeomorphism h.

Now suppose $k > 0$. Then some component J of $S_1 \cap S_2$ bounds a 2-cell $D \subset S_2$ with Int $D \cap S_1 = \emptyset$. Let E' and E'' be the 2-cells in S_1 bounded by J and let $S'_1 = D \cup E'$, $S''_1 = D \cup E''$. At least one of S'_1, S''_1 is nonseparating. This follows from the facts that a 2-sphere in Int M separates M if and only if its intersection number with every loop in M is even, and that the intersection number of any loop with S_1 is congruent, mod 2, to the sum of its intersection numbers with S'_1

and with S''_1. So suppose S'_1 is nonseparating. Since S'_1 can be moved slightly so as to miss S_1 and to meet S_2 in fewer than k components, two applications of induction will complete the proof. Induction applies unless one of $M - S_1$, $M - S'_1$ is orientable and the other is not. If $M - S'_1$ is orientable, then $M - (S'_1 \cup S''_1) = M - (S_1 \cup D)$ is connected; otherwise $M - S_1$ is obtained by identifying two (orientable) components of $M - (S'_1 \cup S''_1)$ along D and is, therefore, orientable. Thus S''_1 is nonseparating. Furthermore $M - S''_1$ is nonorientable; otherwise, $M - D = (M - S'_1) \cup (M - S''_1)$ is the union of two orientable submanifolds intersecting in the connected set $M - (S'_1 \cup S''_1)$ and is thus orientable. This makes $M - S_2$ orientable contradicting hypothesis. So, as above, induction applies with S''_1 replacing S'_1 in its application. The case $M - S_1$ orientable and $M - S'_1$ nonorientable is handled similarly completing the proof.

We need the following variation of 3.18.

3.19. **Lemma**. *Suppose the 3-manifold* M *contains a nonseparating 2-sphere* $S_1 \subset \text{Int } M$ *such that every 2-sphere in* $\text{Int } M - S_1$ *separates* $\text{Int } M - S_1$. *Then given any nonseparating 2-sphere* $S_2 \subset \text{Int } M$, *there is a homeomorphism* $h: M \to M$ *such that* $h(S_1) = S_2$.

Proof. By inducting on the number of components of $S_1 \cap S_2$ and using arguments similar to those used in 3.18, we show that $M - S_1$ is orientable if and only if $M - S_2$ is orientable. The conclusion then follows from 3.18.

3.20. **Lemma**. *If* $M = M_1 \# M_2$ *and each 2-sphere in* $\text{Int } M_i$ $(i = 1,2)$ *separates* M_i, *then each 2-sphere in* Int M *separates* M.

Proof. Choose a 2-sphere $T \subset \text{Int } M$ which realizes the splitting $M = M_1 \# M_2$. For an arbitrary 2-sphere $S \subset \text{Int } M$, one shows that S separates M by inducting on the number of components of $S \cap T$.

A prime factorization $M = M_1 \# \cdots \# M_n$ of a 3-manifold M is *normal* provided that some M_i is $S^2 \times S^1$ only if M is orientable. By 3.17 any prime factorization can be replaced by a normal one. We have

3.21. THEOREM. *Let* $M = M_1 \# \cdots \# M_n = M_1^* \# \cdots \# M_{n^*}^*$ *be two normal, prime factorizations of a compact 3-manifold* M. *Then* $n = n^*$ *and (after reordering)* M_i *is homeomorphic to* M_i^* *(orientation-preserving homeomorphic in the oriented category).*

PROOF. Since the number of 3-cell factors in any prime factorization of M is the number of 2-sphere boundary components of M (cf. 3.7), we may assume $M = \hat{M}$. We assume $n \leq n^*$ and induct on n. The case $n = 1$ is trivial; so assume $n > 1$.

Suppose some M_i (say M_n) is $S^2 \times S^1$. By normality M is orientable. By 3.20 some M_i^* (say $M_{n^*}^*$) is not irreducible. By 3.13, the only prime, orientable, nonirreducible 3-manifold is $S^2 \times S^1$; so $M_{n^*}^* = S^2 \times S^1$. By 3.16 there exist nonseparating 2-spheres S, S^* in Int M with $M_1 \# \cdots \# M_{n-1} = (M - S \times (-1,1))\hat{}$, and $M_1^* \# \cdots \# M_{n^*-1}^* = (M - S^* \times (-1,1))\hat{}$. 3.18 applies to show that

$$M_1 \# \cdots \# M_{n-1} = M_1^* \# \cdots \# M_{n^*-1}^* ,$$

and the conclusion follows by induction.

Next, suppose that some M_i is the nonorientable 2-sphere bundle over S^1, which we denote by P. The argument follows as above: 3.13, 3.19 and normality force some M_i^* to be P. 3.16 and 3.18 allow us to cancel a factor of P from both factorizations. The case in which the hypothesis of 3.18 are not obviously satisfied is handled by 3.19.

Now we suppose that each M_i is irreducible. By 3.20 each 2-sphere in Int M separates M; in particular each M_i^* is irreducible. Now there is a 2-sphere $S \subset \text{Int } M$ such that if we denote the closures of the components of $M - S$ by U and V we have $M_1 \# \cdots \# M_{n-1} = \hat{U}$, $M_n = \hat{V}$. Similarly there exist pairwise disjoint 2-spheres $T_1, \cdots, T_{n^*-1}$ in Int M such that if we denote by $W_1, \cdots, W_{n^*}$ the closures of the components of $M - \cup T_i$, then $\hat{W}_i = M_i^*$ $(i = 1, \cdots, n^*)$. Of all such choices for $\{T_1, \cdots, T_{n^*-1}\}$ in general position with respect to S, suppose we have chosen one for which $S \cap \cup T_i$ has a minimal number of components. We claim that $S \cap \cup T_i = \emptyset$. If not, there is a 2-cell $D \subset S$ such that $\partial D \subset T_i$ for some i and $\text{Int}\, D \cap \cup T_i = \emptyset$. Let E' and E'' be the 2-cells in T_i bounded by ∂D. Then $D \subset W_j$ for some j. Now $D \cup E'$ and $D \cup E''$ bound 3-cells B', B'' in $\hat{W}_j$. One of these must contain the other; otherwise $\hat{W}_j = S^3$. So suppose $B' \subset B''$. Let $T'_i = D \cup E''$ and $T'_k = T_k$ for $k \neq i$. If we denote the closures of the components of $M - \cup T'_k$ by $W'_1 \cdots W'_{n*}$ we see that $\hat{W}'_k = \hat{W}_k$ for all k. Since $\{T'_k\}$ can be isotoped so as to meet S in few components, we have contradicted our minimality assumption. Hence $S \cap \cup T_i = \emptyset$. Thus $S \subset W_{n*}$ (say). Either $V \subset W_{n*}$ or $U \subset W_{n*}$. In the latter case, since $\hat{U} = M_1 \# \cdots \# M_{n-1}$, we must have $n = 2$; so by re-indexing we may suppose that $V \subset W_{n*}$. Hence $M_n = \hat{V} = \hat{W}_{n*} = M^*_{n*}$ and $M_1 \# \cdots \# M_{n-1} = \hat{U} = \hat{W}_1 \# \cdots \# \hat{W}_{n*-1} = M^*_1 \# \cdots \# M^*_{n*-1}$ and the conclusion follows by induction.

We now provide examples promised at the beginning of the chapter with

3.22. EXAMPLE. *Let* M *be any oriented, prime 3-manifold which admits no orientation reversing self homeomorphism. Then* $M \# M$ *is not homeomorphic to* $M \# (-M)$.

PROOF. A homeomorphism $h: M \# M \to M \# (-M)$ would be orientation-preserving as a homeomorphism to one of $M \# (-M)$ or $-(M \# (-M)) = (-M) \# M$. In either case, uniqueness (3.21) in the oriented category gives an orientation-preserving homeomorphism of M to $-M$ contradicting hypothesis.

To see that the hypothesis of 3.22 is not vacuous we note that a lens space $L_{p,q}$ admits an orientation-reversing self homeomorphism if and only if $q^2 \equiv -1 \mod p$. The sufficiency of this condition can be established by construction. The necessity is more involved; however the following homotopy classification of lens spaces is sufficient to provide the needed examples.

3.23. LEMMA. *Suppose* $(p,q) = (p,q') = 1$. *Then there is a homotopy equivalence* $h: L_{p,q} \to L_{p,q'}$ *which*

(i) *preserves orientation if and only if* qq' *is a quadratic residue* mod p *or*

(ii) *reverses orientation if and only if* $-qq'$ *is a quadratic residue* mod p.

PROOF. We define the positive orientation on $L_{p,q}$ as follows. Let (V_1, V_2) be the standard Heegaard splitting of $L_{p,q}$. Then we have a standard pair α, β of oriented loops in ∂V_1 which meet transversely in a single point and with β the boundary of an oriented 2-cell D in V_1. The positive orientation on $L_{p,q}$ is prescribed by requiring that the oriented intersection number, $\alpha \cdot D$, of α with D be +1.

Now $\mu = \alpha^p \beta^q$ bounds a 2-cell in V_2, and for $sp - rq = 1$, $\lambda = \alpha^r \beta^s$ is a longitude for V_2. Note that

$$\alpha = (\alpha^p \beta^q)^s (\alpha^r \beta^s)^{-q} = \mu^s \lambda^{-q}$$

and

$$\beta = (\alpha^p \beta^q)^{-r} (\alpha^r \beta^s)^p = \mu^{-r} \lambda^p .$$

Thus, in $L_{p,q}$, $\alpha \cong \lambda^{-q}$, $\beta \cong \lambda^p$, $\lambda \cong \alpha^r$ ($\cong$ means homotopic). From this it follows that D is a cycle modulo p and thus represents an element $b \in H_2(L_{p,q}; Z_p) \cong Z_p$. Also D represents an element $d \in H_2(L_{p,q}, V_2) \cong Z$, α represents a generator, a, for $H_1(L_{p,q}; Z_p) \cong Z_p$, and λ represents a generator, c, for $H_1(V_2) \cong Z$.

Now let $L = L_{p,q}$ and suppose that $L' = L_{p,q'}$ is similarly described, with primes added to denote the corresponding quantities, and that $h: L \to L'$ is a homotopy equivalence. We may assume that $h(V_2) \subset V'_2$; for we can require that a core of V'_1 does not meet $h(V_2)$ and then deform the complement of a small neighborhood of this core into V'_2.

From the diagram

$$\begin{array}{ccc} 0 & & 0 \\ \downarrow & & \downarrow \\ H_2(L,V_2) & \xrightarrow{\quad h_* \quad} & H_2(L',V_2') \\ \partial \downarrow & & \downarrow \partial \\ H_1(V_2) & \xrightarrow{\quad (h|V_2)_* \quad} & H_1(V_2') \end{array}$$

and the facts that $\partial(d) = pc$, $\partial(d') = pc'$, we see that if $(h|V_2)_*(c) = kc'$, then $h_*(d) = kd'$.

By naturality of the universal coefficient sequence and exact sequences of the appropriate pairs we have commutative diagrams

$$\begin{array}{ccccc} H_2(L,V_2) \otimes Z_p & \xrightarrow{\cong} & H_2(L,V_2;Z_p) & \xleftarrow{\cong} & H_2(L;Z_p) \\ \downarrow h_* \otimes 1 & & \downarrow h_* & & \downarrow h_* \\ H_2(L',V_2') \otimes Z_p & \xrightarrow{\cong} & H_2(L',V_2';Z_p) & \xleftarrow{\cong} & H_2(L';Z_p) \end{array}$$

$$\begin{array}{ccccc} H_1(V_2) \otimes Z_p & \xrightarrow{\cong} & H_1(V_2;Z_p) & \xrightarrow{\cong} & H_1(L;Z_p) \\ \downarrow (h|V_2)_* \otimes 1 & & \downarrow (h|V_2)_* & & \downarrow h_* \\ H_1(V_2') \otimes Z_p & \xrightarrow{\cong} & H_1(V_2';Z_p) & \xrightarrow{\cong} & H_1(L';Z_p) \end{array}$$

From the first we conclude that $h_*(b) = kb'$. From the second, and the fact that $c \otimes 1 \mapsto ra$ and $c' \otimes 1 \mapsto r'a'$, we conclude that $h_*(ra) = kr'a'$. Since $rq \equiv -1(p)$, we have $h_*(a) = -kr'qa'$. Now $a \cdot b = +1$; so $\deg h = h_*(a \cdot b) = k^2 r'q(a' \cdot b') = -k^2 r'q$. Thus $qq' \equiv -k^2 q^2 r'q' \deg h \equiv (kq)^2 \deg h \ (p)$. This completes the proof of the necessity. The sufficiency can be established by construction. We leave the details to the reader.

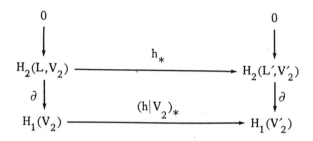

CHAPTER 4
THE LOOP AND SPHERE THEOREMS

In 1910, M. Dehn gave a "proof" [17] of a theorem which has become known as

4.1. DEHN'S LEMMA. *Suppose* M *is a* 3-*manifold and* $f: B^2 \to M$ *is a map such that for some neighborhood* A *of* ∂B^2 *in* B^2 $f|A$ *is an embedding and* $f^{-1}(f(A)) = A$. *Then* $f|\partial B^2$ *extends to an embedding* $g: B^2 \to M$.

However, Dehn's proof contained a serious gap and the problem remained open until the mid 1950's when C. D. Papakyriakopoulos [80], [81] gave a valid proof of Dehn's lemma together with two other theorems (stated below) which he called the loop theorem and the sphere theorem. Since then, several papers [20], [92], [95], [104], [110], [111] have appeared which provide somewhat simpler proofs, generalizations, and in some cases elimination of unnecessary hypothesis (e.g. orientability in the loop theorem, and a certain technical condition in the sphere theorem). It is interesting to note, in retrospect, that Dehn's lemma turns out to be the simplest and least useful of these theorems. The loop and sphere theorems provide a link between algebra (homotopy theory) and geometry and are at the basis of almost all results in 3-manifold topology.

The following theorem, due to J. Stallings [95], combines the original statement of the loop theorem with Dehn's lemma.

4.2. THE LOOP THEOREM. *Let* M *be any* 3-*manifold and* F *a connected* 2-*manifold in* ∂M. *If* N *is a normal subgroup of* $\pi_1(F)$ *and if* $\ker(\pi_1(F) \to \pi_1(M)) - N \neq \emptyset$, *then there is a proper embedding* $g: (B^2, \partial B^2) \to (M, F)$ *such that* $[g|\partial B^2] \notin N$.

Noting that for $n \geq 2$ $\pi_1(x)$ acts on $\pi_n(x)$, a subgroup A of $\pi_n(x)$ is called $\pi_1(x)$-*invariant* if it is left setwise fixed under the action of every element of $\pi_1(x)$.

4.3. THE SPHERE THEOREM. *If* M *is any orientable 3-manifold and if* N *is any* $\pi_1(M)$-*invariant subgroup of* $\pi_2(M)$ *with* $\pi_2(M) - N \neq \emptyset$, *then there is an embedding* $g: S^2 \to M$ *such that* $[g] \notin N$.

4.4. EXERCISE. *Deduce* 4.1 *as a corollary to* 4.2. (Hint: Let R be a regular neighborhood of $f(\partial B^2)$ in M and apply 4.2 to $\overline{M - R}$.)

The loop theorem (with $N = 1$) yields the following. If $K = \ker(\pi_1(F) \to \pi_1(M)) \neq 1$, then some nontrivial element of K is represented by a simple loop. This should be contrasted with the fact that "lots" of (normal) subgroups of $\pi_1(F)$ don't contain any such elements.

Observe that the sphere theorem is false, as stated, for nonorientable 3-manifolds. For example $\pi_2(P^2 \times S^1) = Z$, but $P^2 \times S^1$ is irreducible (3.11, 3.13); hence, any embedding of S^2 into $P^2 \times S^1$ is nullhomotopic. An extension of the sphere theorem to the nonorientable case is given in 4.12.

In each of 4.1, 4.2, and 4.3, the hypothesis gives a (singular) map $f: F \to M (F = B^2$ or $S^2)$ which satisfies certain conditions ($[f] \notin N$ in 4.2 and 4.3) while the conclusion promises a nonsingular map $g: F \to M$ satisfying the same conditions.

Initially, efforts were made to prove Dehn's lemma by reducing the given (general position) map $f: B^2 \to M$ to an embedding by "cut and paste" techniques. While these techniques (described below) play a role in the present proofs, they are not sufficient: Johansson [53] gave numerous examples to illustrate this in 1935 (see exercise 4.8 below).

The essence of Papakyriakopoulos' approach to these problems is to construct a factorization

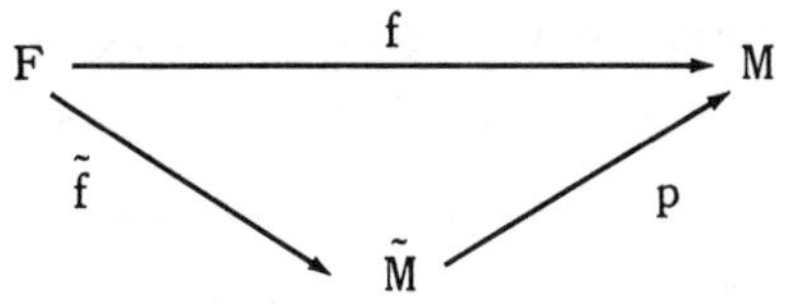

where the corresponding problem for $\tilde{M}$ has a "solution" $\tilde{g}$ (for purely algebraic reasons) and where the composition $p \circ \tilde{g}$ can, in fact, be reduced by cut and paste techniques. He constructs $\tilde{M}$ from a "*tower of covering spaces.*" This idea is common to all variations of these theorems and is useful elsewhere in the theory.

The term 'essence' in the preceding paragraph was chosen deliberately. It applies most accurately to the loop theorem (particularly Stalling's version). The sphere theorem is a bit more complicated.

Double Curve Surgery

We first describe cutting and pasting (surgery). Let f be a compact 2-manifold (with or without boundary) and M be a 3-manifold. Let $f:(F, \partial F) \to (M, \partial M)$ be a general position map. We use the notation of 1.11:

$$S_i(f) = \{x \in S(f) : \# f^{-1}(f(x)) = i\}, \quad \Sigma_i(f) = f(S_i(f)), \text{ etc.}$$

Recall that $S_i(f) = \emptyset$ for $i \geq 4$, that $S_1(f)$ and $S_3(f)$ are finite, and that $f|F - S_1(F)$ is an immersion. Pick a double point, y, of f (a point of $\Sigma_2(f)$). A neighborhood of y in $\Sigma_2(f)$ is an arc in which two sheets of $f(F)$ intersect transversely. By choosing an orientation for this arc, we have locally defined a path in $\Sigma(f)$. By passing to overlapping neighborhoods we may continue this path as long as we do not encounter a branch point or a point of ∂M (continuation across a triple point causes no difficulty by transversality). By compactness, a maximal continuation of this path must either

(i) close up, or

(ii) end in a branch point or a point of ∂M.

In case (i) we have constructed a *closed double curve of* f: a map $\alpha : S^1 \to \Sigma(f)$ with $S(\alpha)$ finite and $\Sigma(\alpha) \subset \Sigma_3(f)$. In case (ii) we continue in the opposite direction and must eventually encounter another branch point or boundary point. In this case we have constructed a *double arc of* f: a map $\alpha : I \to \Sigma(f)$ with $S(\alpha)$ finite, $\Sigma(\alpha) \subset \Sigma_1(f) \cup \Sigma_3(f)$, $\alpha(\text{Int } I) \subset \text{Int } M - \Sigma_1(f)$, and $\alpha(\partial I) \subset \partial M \cup \Sigma_1(f)$. We use the term *double curve of* f for either a closed double curve or a double arc. We have proved

4.5. LEMMA. *If* F *is a compact 2-manifold,* M *is a 3-manifold and* $f : (F, \partial F) \to (M, \partial M)$ *is a general position map, then there is a finite collection* $\alpha_1, \cdots, \alpha_n$ *of double curves of* f *such that* image $\alpha_i \cap$ image $\alpha_j \subset \Sigma_1(f) \cup \Sigma_3(f)$ *if* $i \neq j$ *and* $\Sigma(f) = \bigcup_{i=1}^{n}$ image α_i. *The* α_i *are unique up to* parametrization.

We say that a double curve α is *simple* if α is an embedding and image $\alpha \cap \Sigma_1(f) = \emptyset$ (the second condition is automatically satisfied for closed double curves). Since $f|F - S_1(f)$ is an immersion, we see that for a simple double curve α

$$f \,|\, f^{-1}(\text{image } \alpha) : f^{-1}(\text{image } \alpha) \to \text{image } \alpha$$

is a covering map. So if α is a simple double arc, α is a proper embedding and $f^{-1}(\alpha(I))$ has two components, each of which is properly embedded in F and maps homeomorphically, via f, to $\alpha(I)$. If α is a simple closed double curve, then either $f^{-1}(\alpha(S^1))$ has two components each projecting homeomorphically to $\alpha(S^1)$ or $f^{-1}(\alpha(S^1))$ has one component which doubly covers $\alpha(S^1)$.

We define the *complexity* of a general position map $f : (F, \partial F) \to (M, \partial M)$ to be the pair $c(f) = (t(f), d(f))$ where $t(f)$ is the number of triple points of $f (= \#(\Sigma_3(f)))$ and $d(f)$ is the number of double curves of f. We order complexities lexicographically; i.e. $c(f_1) < c(f)$ if either $t(f_1) < t(f)$ or $t(f_1) = t(f)$ and $d(f_1) < d(f)$.

4.6. LEMMA. *Let* F *be a compact 2-manifold,* M *be a 3-manifold, and* $f:(F,\partial F)\to(M,\partial M)$ *a general position map. Suppose* f *has a simple closed double curve* α *such that each component of* $f^{-1}(\alpha(S^1))$ *bounds a 2-cell in* F*. Then given any neighborhood* U *of* $\alpha(S^1)$ *in* M*, there is a general position map* $f_1:(F,\partial F)\to(M,\partial M)$ *satisfying:*

(i) $c(f_1)<c(f)$,

(ii) $f_1(F)\subset f(F)\cup U$, *and*

(iii) f_1 *agrees with* f *off a (preassigned) neighborhood of the 2-cell(s) in* F *bounded by* $f^{-1}(\alpha(S^1))$.

Finally, if $f^{-1}(\alpha(S^1))$ *is connected, then* α *is orientation reversing (so* M *is nonorientable).*

PROOF. We consider the cases:

Case 1. $f^{-1}(\alpha(S^1))$ has two components J_1 and J_2.
Case 2. $J=f^{-1}(\alpha(S^1))$ is connected.

In case 1 we have, by hypothesis, 2-cells $D_1,D_2\subset F$ with $\partial D_i=J_i$. We further divide into 1a: $D_1\cap D_2=\emptyset$ and 1b: $D_1\subset \operatorname{Int} D_2$.

Consider case 1a. Since $f|J_1$ is a homeomorphism, α is null homotopic; thus α is orientation preserving. So a regular neighborhood T of $\alpha(S^1)$ is a solid torus in Int M. We may choose T so that $f^{-1}(T)$ is the union of two disjoint annuli A_1, A_2 with $J_i\subset \operatorname{Int} A_i$ and such that $f|A_i:A_i\to T$ is a proper embedding. Let $D'_i=D_i-\operatorname{Int} A_i$, $D''_i=D_i\cup A_i$. There are disjoint annuli $B_1,B_2\subset\partial T$ with $\partial B_1=f(\partial D''_1\cup\partial D'_2)$ and $\partial B_2=f(\partial D''_2\cup\partial D'_1)$. We define f_1 by putting $f_1|F-(D''_1\cup D''_2)=f|F-(D''_1\cup D''_2)$, extending f_1 to take A_i homeomorphically to B_i, and further extending so that $f_1(D'_1)\subset f(D'_2)$ and $f_1(D'_2)\subset f(D'_1)$. We may take $f_1|D'_1=(f|D'_2)\circ h$ for a properly chosen homeomorphism $h:D'_1\to D'_2$ and vice versa.

Clearly (ii) and (iii) are satisfied (by taking T small). For (i) we note that $\Sigma_3(f_1)\subset\Sigma_3(f)-T$. Hence, if $\alpha(S^1)\cap\Sigma_3(f)\neq\emptyset$, $t(f_1)<t(f)$. If $\alpha(S^1)\cap\Sigma_3(f)=\emptyset$, $t(f_1)=t(f)$ and $d(f_1)<d(f)$. In either case

$c(f_1) < c(f)$. Note that in the first case we may have $d(f_1) > d(f)$ – a double curve of f which meets $\alpha(S^1)$ may be split into two double curves of f_1.

Case 1b follows similarly using the fact that in this case α is also orientation preserving.

For case 2 we let T be a regular neighborhood of $\alpha(S^1)$ chosen so that T does not meet any double curve of f which is disjoint from α and such that $A = f^{-1}(T)$ is a regular neighborhood of J in F. Since J bounds a 2-cell D in F, A is an annulus. By choice of T, $S(f|A) = J$; in particular $f|\partial A : \partial A \to \partial T$ is an embedding. Now some 2-cell $E \subset T$ cuts T into a product $B^2 \times I$ and cuts $f(F) \cap T$ into a subproduct $X \times I$ where X is the union of two arcs properly embedded in B^2 which meet transversely in a single point. By investigating the eight essentially different ways in which $B^2 \times 0$ could be identified with $B^2 \times 1$ (and $X \times 0$ with $X \times 1$) to yield T, we see that only two of these yield $T \cap f(F)$ as the image of a general position map of an annulus as described above. These two yield T as a solid Klein bottle. Thus α is orientation reversing. Let $D' = D - \text{Int } A$, $D'' = D \cup A$. Since $f|\partial D'$ is an embedding, it is orientation preserving in M. Now ∂T is 2-sided in M so $f(\partial D')$ is 2-sided in ∂T. Thus $f(\partial D')$ and $f(\partial D'')$ cobound an annulus $B \subset \partial T$. We obtain the desired map f_1 with

$$f_1|F - D'' = f|F - D'', \quad f_1(A) = B, \quad \text{and} \quad f_1(D') \subset f(D') .$$

4.7. LEMMA. *Let* M *be a 3-manifold and* $f : (B^2, \partial B^2) \to (M, \partial M)$ *a general position map. Suppose* f *has a simple double arc* α. *Then given any neighborhood* U *of* $\alpha(I)$, *there are general position maps*

$$f_1, f_2 : (B^2, \partial B^2) \to (M, \partial M)$$

satisfying

(i) $c(f_i) < c(f)$ $(i = 1,2)$,

(ii) $f_i(B^2) \subset f(B^2) \cup U$, *and*

(iii) $[f|\partial B^2]$ *lies in the smallest normal subgroup of* $\pi_1(\partial M)$ *containing* $[f_1|\partial B^2]$ *and* $[f_2|\partial B^2]$.

PROOF. There are two liftings $\alpha_1, \alpha_2 : (I, \partial I) \to (B^2, \partial B^2)$ of α, and essentially two possibilities for their relative orientation on B^2 as shown below.

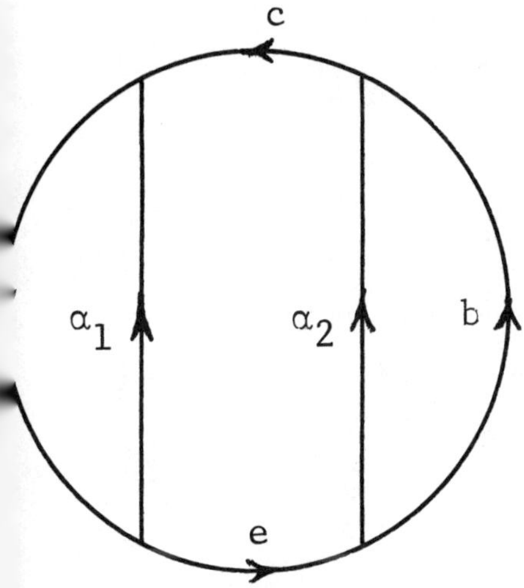

Case 1

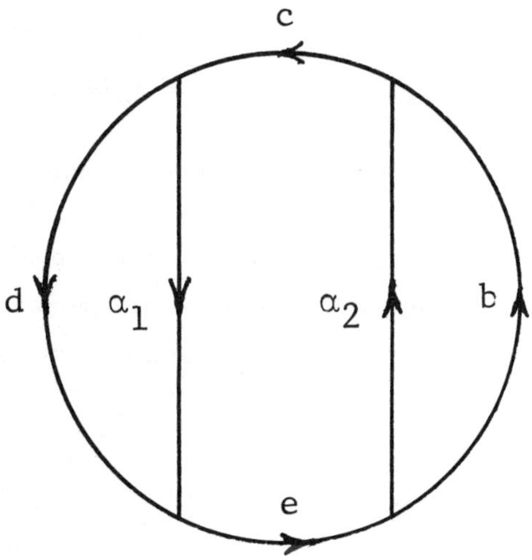

Case 2

Copying the proof of the preceding lemma, we get two maps

$$f_1, f_2 : (B^2, \partial B^2) \to (M, \partial M)$$

satisfying (i) and (ii) where f_1 omits the "middle" part of $f(B^2)$ and f_2 reverses it. Writing ∂B^2 as a path product, $b\,c\,d\,e$, as shown and putting $\beta = f \circ b$, $\gamma = f \circ c$, $\delta = f \circ d$, $\varepsilon = f \circ e$, we have in case 1; $f_1 | \partial B^2 \cong B\delta$ and $f_2 | \partial B^2 \cong \beta\gamma^{-1}\delta\varepsilon^{-1}$ ($\cong$ means homotopic in ∂M). Thus

$$\begin{aligned}[f|\partial B^2] &= [\beta\gamma\delta\varepsilon] = [\beta\delta\varepsilon^{-1}\varepsilon\delta^{-1}\gamma\beta^{-1}\beta\delta\varepsilon] \\ &= [\beta\delta][\varepsilon]^{-1}[\beta\gamma^{-1}\delta\varepsilon^{-1}]^{-1}[\beta\delta][\varepsilon] \\ &= [f_1|\partial B^2][\varepsilon]^{-1}[f_2|\partial B^2]^{-1}[f_1|\partial B^2][\varepsilon].\end{aligned}$$

This gives (iii). We note that in the above computation, all paths enclosed in brackets are loops based at $\beta(0)$.

In case 2, $f_1|\partial B^2 \cong \beta\delta^{-1}$ and $f_2|\partial B^2 \cong \beta\,\epsilon\,\delta\gamma$. Thus

$$
\begin{aligned}
[f|\partial B^2] &= [\beta\gamma\delta\varepsilon] = [\beta\delta^{-1}\varepsilon^{-1}\delta^{-1}\delta\beta^{-1}\beta\varepsilon\delta\gamma\delta\varepsilon] \\
&= [\beta\delta^{-1}][\delta\varepsilon]^{-1}[\beta\delta^{-1}]^{-1}[\beta\varepsilon\delta\gamma][\delta\varepsilon] \\
&= [f_1|\partial B^2][\delta\varepsilon]^{-1}[f_1|\partial B^2]^{-1}[f_2|\partial B^2][\delta\varepsilon],
\end{aligned}
$$

and we have completed the proof.

4.8. EXERCISE. *Show that the diagram below describes the singular set,* S(f), *of a general position map* f *of* B^2 *into a* 3-*manifold* M. *Observe that* $c(f) = (2;1)$ *and that* f *has no simple double curve.*

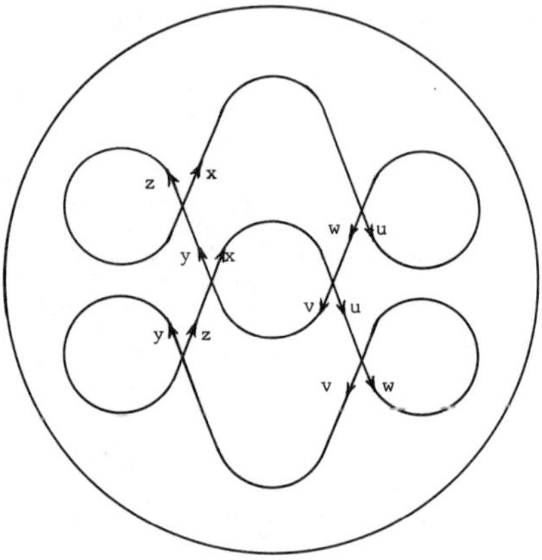

REMARKS. The preceding diagram describes S(f) and indicates how the points of S(f) are identified by f. This clearly defines $f(B^2)$, in the abstract, as an identification space. One must show that $f(B^2)$ can be embedded in a 3-manifold M so as to make f a general position map. This can be done by constructing M by a handle decomposition: taking disjoint 3-cells (to be neighborhoods of the triple points), adding 1-handles to form a neighborhood of the double curves, and so on. Clearly, this procedure will not work for an arbitrary "diagram." Conditions for a diagram to be "realizable" and other examples are given in [53].

4.9. LEMMA. *Let* M *be a compact* 3-*manifold with nonempty boundary. If some component of* ∂M *is not a* 2-*sphere, then* M *has a* (*connected*) *double covering.*

PROOF. We must show that $\pi_1(M)$ contains a subgroup of index two. Since any map $\pi_1(M) \to Z_2$ factors through $H_1(M; Z_2)$ (via the composition of the Hurewicz homomorphism with reduction mod 2), it suffices (and is necessary) to show $H_1(M; Z_2) \neq 0$.

So suppose $H_1(M; Z_2) = 0$. Then $H_2(M, \partial M; Z_2) = H^1(M; Z_2) = 0$. From the exact sequence

$$\to H_2(M, \partial M; Z_2) \to H_1(\partial M; Z_2) \to H_1(M; Z_2) \to$$

it follows that $H_1(\partial M; Z_2) = 0$. This gives the contradiction that all components of ∂M are 2-spheres.

Proof of the Loop Theorem

The following version clearly implies 4.2 and gives more information about the embedding g promised by the conclusion.

4.10. THEOREM. *Let* M *be any* 3-*manifold,* F *be any connected* 2-*manifold in* ∂M, *and* N *be a normal subgroup of* $\pi_1(F)$. *Suppose* $f : (B^2, \partial B^2) \to (M, F)$ *is a general position map such that* $[f|\partial B^2] \notin N$. *Then given any neighborhood* U *of* $\Sigma(f)$, *there is an embedding* $g : (B^2, \partial B^2) \to (M, F)$ *satisfying*

(i) $[g|\partial B^2] \notin N$, *and*

(ii) $g(B^2) \subset f(B^2) \cup U$.

PROOF. Let V_0 be a regular neighborhood of $f(B^2)$ in M such that $F_0 = V_0 \cap F$ is a regular neighborhood of $f(\partial B^2)$ in F. Let $N_0 = i_{0*}^{-1}(N) < \pi_1(F_0)$, and let $f_0 = f$; so $[f_0|\partial B^2] \notin N_0$.

Let $p_1 : M_1 \to V_0$ be a (connected) 2-sheeted covering (if one exists). Since $\pi_1(B^2) = 1$, f_0 lifts to a map $f_1 : (B^2, \partial B^2) \to (M_1, p_1^{-1}(F_0))$ such that $p_1 \circ f_1 = f_0$. Let V_1 be a regular neighborhood of $f_1(B^2)$ in M_1 such that $F_1 = V_1 \cap p_1^{-1}(F_0)$ is a regular neighborhood of $f_1(\partial B^2)$ in $p_1^{-1}(F_0)$. Let $N_1 = (p_1 \circ i_1)_*^{-1}(N_0) < \pi_1(F_1)$; note that $[f_1|\partial B^2] \notin N_1$. We continue this construction as long as possible. After n steps we have a *tower of height* n:

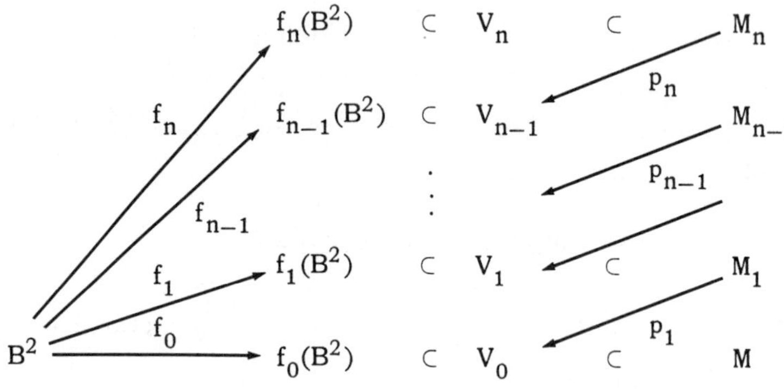

A tower of height n can be extended to one of height (n+1) provided that V_n has a 2-sheeted covering space.

We want one further condition: that the maps $f_1, \cdots, f_n$ be simultaneously simplicial with respect a fixed triangulation of B^2. This can be accomplished as follows. Choose triangulations T of B^2 and K of M with respect to which f is simplicial and such that F underlies some subcomplex of K. Then $f(B^2) = |L|$ for some $L \subset K$. Let K' mod L be the subdivision of K obtained by adding only the barycenters of simplexes of K − L. Now L is a subcomplex of K' mod L; so we may iterate to construct K'' mod L. Using Theorem 1.6 (compare with Corollary 1.7), one can show that $N(L, K'' \bmod L)$ is a regular neighborhood of $f(B^2)$ which we choose to be V_0. Now K'' mod L restricts to a triangulation of V_0 which we may lift to get a triangulation K_1 of M_1. Then f_1 is simplicial as a map from T to K_1; so we may iterate the construction to obtain the desired result.

Let $X(f_i) = \{(\sigma, \tau) \in T \times T : \sigma \neq \tau$ and

$$f_i(\text{Int}\,\sigma) \cap f_i(\text{Int}\,\tau) \neq \emptyset\} ;$$

so $X(f_i)$ is a finite set of pairs of simplexes of T. Clearly $X(f_{i+1}) \subseteq X(f_i)$. We claim that $X(f_{i+1}) \neq X(f_i)$; for if not, it follows that $p_{i+1}|f_{i+1}(B^2)$ is one to one. But $i_*: \pi_1(f_j(B^2)) \to \pi_1(V_j)$ is an isomorphism for all j; hence

$$(i \circ (p_{i+1}|f_{i+1}(B^2)))_* : \pi_1(f_{i+1}(B^2)) \to \pi_1(V_i)$$

is an isomorphism. This contradicts the fact that $p_{i+1\,*}\, \pi_1(V_{i+1})$ has index two in $\pi_1(V_i)$.

So we see that the tower construction cannot be continued to a height greater than $\# X(f_0)$, and that we thus have a tower of some finite height n such that V_n has no (connected) double cover. By 4.9 each component of ∂V^n is a 2-sphere. Thus F_n is a planar surface and $[f_n|\partial B^2] \notin N_n$. Now the fundamental group of a planar surface is normally generated by its boundary components. Since N_n is a proper normal subgroup of $\pi_1(F_n)$, there is a component J of ∂F_n such that $[J] \notin N_n$. J bounds a 2-cell D in ∂V_n with $D \cap F_n = J$.

Keeping in mind part (ii) of the conclusion, let $U_0 = U \cap V_0$, $U_1 = p_1^{-1}(U_0) \cap V_1$, etc. By uniqueness of regular neighborhoods, we may assume that V_n is constructed as follows. Let R be a regular neighborhood of $\Sigma(f_n)$ in M_n. We may suppose $R \subset p_n^{-1}(U_{n-1})$. The components $C_1, \cdots, C_k$ of $f_n(B^2) - \text{Int } R$ are (nonsingular) planar surfaces. Choose product neighborhoods $C_j \times [-1,1]$ of $C_j (C_j = C_j \times 0)$ in $M_n - \text{Int } R$ which meet ∂R regularly, and let $V_n = R \cup C_1 \times [-1,1] \cup \cdots \cup C_k \times [-1,1]$. For no integer i do we have both $(C_i \times (-1)) \cap D \neq \emptyset$; and $(C_i \times 1) \cap D \neq \emptyset$; otherwise we could join, in $C_i \times [-1,1]$, the ends of some arc in D to obtain a simple closed curve which has intersection number 1 with the relative 2-cycle represented by $f_n(B^2)$. This curve represents an element of infinite order in $H_1(V_n)$. Thus $H_1(V_n; Z_2) \neq 0$ and V_n has a 2-sheeted cover contradicting the choice of n.

By the preceding remarks we can adjust D by an isotopy so as to lie in $R \cup f_n(B^2)$. This gives an embedding $g_n : (B^2, \partial B^2) \to (V_n, F_n)$ such that $[g_n|\partial B^2] \notin N_n$ and $g_n(B^2) \subset f_n(B^2) \cup U_n$. We may further assume (after modifying g_n in $g_n^{-1}(R)$) that $p_n \circ g_n : (B^2, \partial B^2) \to (V_{n-1}, F_{n-1})$ is a general position map. Since P_n is a double covering map

$$\Sigma_1(p_n \circ g_n) = \Sigma_3(p_n \circ g_n) = \emptyset ;$$

thus every double curve of $p_n \circ g_n$ is simple. Hence we can apply Lemmas 4.6 and 4.7 to replace $p_n \circ g_n$ by an embedding

$$g_{n-1} : (B^2, \partial B^2) \to (V_{n-1}, F_{n-1})$$

such that $g_{n-1}(B^2) \subset f_{n-1}(B^2) \cup U_{n-1}$ and $[g_{n-1}|\partial B^2] \notin N_{n-1}$. Notice that part (iii) of 4.7 is essential in this application. By repeating this process we obtain the desired embedding $g = g_0 : (B^2, \partial B^2) \to (M, F)$.

Proof of the Sphere Theorem

It is tempting to try to adapt the above proof directly (replacing B^2 by S^2, etc.). Most steps go through with little change; one step becomes a major problem: that of finding an embedding $g_n : S^2 \to V_n$ at the top of the tower with $[g_n] \notin N_n$. One cannot expect (as one might, by analogy, hope) to get g_n as the inclusion of some 2-sphere in ∂V_n. This is because $\pi_2(V_n)$ need not be generated as a $\pi_1(V_n)$-module by its boundary components (e.g. we might have $\hat{V}_n = L_{p_1,q_1} \# L_{p_2,q_2}$, $p_1 p_2$ odd. Then V_n has no double cover, but $\pi_2(\hat{V}_n) \neq 0$).

To overcome this difficulty, we use a different tower construction involving universal covers. This complicates the remainder of the argument since we may no longer assume, for example, that $p_n \circ g_n$ has any simple double curve.

We reformulate the sphere theorem as

4.11. THEOREM. *Let* M *be an orientable* 3*-manifold,* N *be a* $\pi_1(M)$*-invariant subgroup of* $\pi_2(M)$*, and* $f : S^2 \to M$ *be a general position map such that* $[f] \notin N$*. Then given any neighborhood* U *of* $\Sigma(f)$*, there is an embedding* $g : S^2 \to M$ *such that*

(i) $[g] \notin N$, *and*

(ii) $g(S^2) \subset f(S^2) \cup U$.

PROOF. If $c(f) = (o, o)$, then f is an embedding. So we proceed by induction. We wish to postpone consideration of branch points; so we assume that $\Sigma_1(f) = \emptyset$ and that the conclusion holds for all maps f' (manifolds M', etc.) such that $c(f') < c(f)$ and $\Sigma_1(f') = \emptyset$. We break the proof into steps.

1. *If* f *has a simple closed double curve,* J, *then the conclusion holds* (for f). For $f^{-1}(J)$ has two components J_1 and J_2; otherwise by 4.6, M is nonorientable. There are two essentially different ways of applying 4.6 (corresponding to choices for the 2-cells in S^2 bounded by J_1 and J_2). Thus we get maps $f_1, f_2 : S^2 \to M$ with $f_i(S^2) \subset f(S^2) \cup U$ and $c(f_i) < c(f)$ $i = 1, 2$. One checks that $[f]$ is contained in the smallest $\pi_1(M)$-invariant subgroup of $\pi_2(M)$ containing $[f_1]$ and $[f_2]$. Thus $[f_1] \notin N$ or $[f_2] \notin N$ and the step is completed by induction.

2. *The tower construction.* Let V_0 be a regular neighborhood of $f(S^2)$ in M. Let $N_0 = i_*^{-1}(N)$ (a $\pi_1(V_0)$-invariant subgroup of $\pi_2(V_0)$). Let $f_0 = f$. The first step of the tower construction is described separately as follows. If $\pi_1(V_0)$ is finite, we stop (with a tower of height zero); otherwise we consider the (noncompact) universal cover $p : \tilde{V}_0 \to V_0$. Let $\tilde{f}_0 : S^2 \to \tilde{V}_0$ be a lifting of f. Now $p^{-1}(f_0(S^2))$ is connected (since $\pi_1(f_0(S^2)) \underset{\to}{\cong} \pi_1(V_0)$) and noncompact. Thus there is a covering translation $\tau \neq 1$ of $\tilde{V}_0$ such that $\tilde{f}_0(S^2) \cap \tau \tilde{f}_0(S^2) \neq \emptyset$. Let $M_1 = \tilde{V}_0/\tau$ be the quotient of $\tilde{V}_0$ by the action of the cyclic group generated by τ and let $q : \tilde{V}_0 \to M_1$ be the covering projection. We have a covering map $p_1 : M_1 \to V_0$ such that $p = p_1 \circ q$. Thus $f_1 = q \circ \tilde{f}_0 : S^2 \to M_1$ is a lifting of f_0. Let V_1 be a regular neighborhood of $f_1(S^2)$ in M_1. Let $N_1 = (p_1 \circ j)_*^{-1}(N_0)$; so N_1 is a $\pi_1(V_1)$-invariant subgroup of $\pi_2(V_1)$ and $[f_1] \notin N_1$.

If $\pi_1(V_1)$ is finite we stop; otherwise, let $p_2 : M_2 \to V_1$ be the universal cover. Choose a lifting $f_2 : S^2 \to M_2$ of f_1 and a regular neighborhoof V_2 of $f_2(S^2)$. Let $N_2 = (p_2 \circ i)_*^{-1}(N_1)$. We continue this construction as long as possible using the universal cover $p_k : M_k \to V_{k-1}$ for $k \geq 2$. The argument given in the loop theorem shows that this process must terminate. Thus we get a tower of some finite height n; where (by choice) $\pi_1(V_n)$ is finite and $\pi_1(V_k)$ is infinite for $k < n$.

3. *If* f_n *is an embedding, then* f_0 *has a simple closed double curve, and the proof is completed by Step 1.* Since f_1 is singular by construction, we have $n \geq 2$. Note that $\pi_1(M_{n-1})$ is trivial if $n > 2$ and is cyclic if $n = 2$. If $H_1(V_{n-1})$ is finite, then each component of ∂V_{n-1} is a 2-sphere and $i_* : \pi_1(V_{n-1}) \to \pi_1(M_{n-1})$ is monic. Hence, $\pi_1(V_{n-1})$ is abelian. Thus, $\pi_1(V_{n-1}) \cong H_1(V_{n-1})$ is finite contrary to definition of n. Thus, $H_1(V_{n-1})$ is infinite.

Let A be the group of covering translations of (M_n, p_n), $\eta : A \to \pi_1(V_{n-1})$ be the natural isomorphism, and $\theta : \pi_1(V_{n-1}) \to H_1(V_{n-1})$ be the Hurewicz map. Put $\phi = \theta \circ \eta : A \to H_1(V_{n-1})$ and let T be the torsion subgroup of $H_1(V_{n-1})$. Then $K = \phi^{-1}(T)$ is a proper subgroup of A. Now $p_n^{-1}(f_{n-1}(S^2)) = \cup\{\sigma f_n(S^2) : \sigma \in A\}$ is connected; so there exist $\sigma \in K$, $\mu \notin K$ such that $\sigma f_n(S^2) \cap \mu f_n(S^2) \neq \emptyset$. Let $\tau = \sigma^{-1}\mu$. Then $f_n(S^2) \cap \tau(f_n(S^2)) \neq \emptyset$ and $\tau \notin K$; thus τ has infinite order.

Let $r > o$ be the largest integer such that $f_n(S^2) \cap \tau^r(f_n(S^2)) \neq \emptyset$, and let J be a component of $f_n(S^2) \cap \tau^r(f_n(S^2))$. We claim that $p_n|J$ is one to one. If not, then there exist $x \in J$, $1 \neq \rho \in A$ such that $\rho(x) \in J$. Let E be a neighborhood of x in $f_n(S^2)$. Now $\rho(E)$ lies in one of the (at most 3) translates of $f_n(S^2)$ which contain $\rho(x)$. We cannot have $\rho(E) \subset f_n(S^2)$; since $f_{n-1} = p_n \circ f_n$ is a general position map. If $\rho(E) \subset \tau^r(f_n(S^2))$, then $\rho = \tau^r$ and $\rho(x) \in f_n(S^2) \cap \rho(\tau^r(f_n(S^2))) = f_n(S^2) \cap \tau^{2r}(f_n(S^2))$ contrary to the choice of r. The third possibility,

$$\rho(E) \not\subset f_n(S^2) \cup \tau^r(f_n(S^2)) ,$$

gives $\rho = \tau^{-r}$ and, hence, the contradiction that

$$x \in f_n(S^2) \cup \tau^{2r}(f_n(S^2)) .$$

Thus $p_n(J)$ is a simple double curve of f_{n-1}. By general position $p_1 \circ \cdots \circ p_n(J)$ is a simple double curve of $f_0 = p_1 \circ \cdots \circ p_n \circ f_n$; otherwise for some y $f_0^{-1}(y)$ contains at least four points. This completes Step 3.

4. *If* f *has no simple closed double curve, then there is a general position map* $f': S^2 \to M$ *such that* $[f'] \notin N$, $f'(S^2) \subset f(S^2) \cup U$, $\Sigma_1(f') = \emptyset$, *and* $t(f') < t(f)$. *Thus the conclusion follows by induction.*

By Step 3, f_n is singular. Since $\pi_1(V_n)$ is finite, the universal cover of $\hat{V}_n$ is a homotopy 3-sphere. Thus $\pi_2(V_n)$ is generated by the components of ∂V_n; therefore, some component S of ∂V_n represents an element of $\pi_2(V_n) - N_n$. As in the loop theorem, we may assume that $V_n = R \cup \cup C_j \times [-1,1]$ where R is a (small) regular neighborhood of $\Sigma(f_n)$ and $C_1, \cdots, C_k$ are the components of $f_n(S^2) - \operatorname{Int} R$. The same proof shows that for no j do both $C_j \times (-1)$ and $C_j \times 1$ meet S. Thus we may adjust the inclusion $S \subset V_n$ by an isotopy to obtain an embedding: $f'_n : S^2 \to V_n$ such that $[f'_n] \notin N_n$ and $f'_n(S^2) \subset \partial R \cup \cup C_j$. Let $f' = p_1 \circ \cdots \circ p_n \circ f'_n : S^2 \to M$. Now, by general position, $p_1 \circ \cdots \circ p_n | \Sigma(f_n)$ is one to one; so we may assume $p_1 \circ \cdots \circ p_n | R$ is one to one.

Let $y \in \Sigma_3(f')$ be a triple point of f'. If $f'^{-1}(y) \cap f'^{-1}_n(R) = \emptyset$, then y is a triple point of f (but $y \notin p_1 \circ \cdots \circ p_n(\Sigma_3(f_n))$). If $f'^{-1}(y) \cap f'^{-1}_n(R) \neq \emptyset$, then this set contains exactly one point. Thus y lies on a double curve of f, and proceeding along this double curve into $p_1 \circ \cdots \circ p_n(R)$ we arrive at a triple point z of f (again $z \notin p_1 \circ \cdots \circ p_n(\Sigma_3(f_n))$). Distinct triple points of f' do not lead to the same triple point of f. This follows from the properties of S – specifically that S does not meet both $C_j \times (-1)$ and $C_j \times 1$ for any j. Thus we have a one-to-one correspondence from the set of triple points of f' into the set of triple points of f which are not images of triple points of f_n. Thus we have $t(f') < t(f)$ unless $t(f_n) = 0$. If $t(f_n) = 0$, then, since f_n has no branch points, each double curve of f_n is simple and closed. By general position the image, under $p_1 \circ \cdots \circ p_n$ of such a double curve would be a simple double curve of f. This completes Step 4.

To complete the proof of the sphere theorem we must consider the case in which f has branch points. In this case we repeat the tower construction of Step 2 and use the arguments of Step 4 to construct a map $f' = p_1 \circ \cdots \circ p_n \circ f'_n$ where $f'_n : S^2 \to V_n$ is an embedding. Since each p_i is an immersion, f' has no branch points and the proof is completed by the special case. We observe that it is necessary to complete the full argument for the special case first since we cannot conclude that $c(f') < c(f)$ if f_n has branch points.

The Projective Plane Theorem

The following generalization of the sphere theorem to nonorientable 3-manifolds is due to Epstein [20].

4.12. THEOREM. *Let* M *be any* 3-*manifold and* N *a* $\pi_1(M)$-*invariant subgroup of* $\pi_2(M)$. *If* $f : S^2 \to M$ *is a general position map such that* $[f] \notin N$ *and* U *is any neighborhood of* $\Sigma(f)$, *then there is a map* $g : S^2 \to M$ *satisfying*

(i) $[g] \notin N$,

(ii) $g(S^2) \subset f(S^2) \cup U$,

(iii) $g : S^2 \to g(S^2)$ *is a covering map, and*

(iv) $g(S^2)$ *is a* 2-*sided submanifold* (2-*sphere or projective plane*) *of* M.

PROOF. We assume M is nonorientable and let $p : \tilde{M} \to M$ be the orientable double cover. Choose a lifting $\tilde{f} : S^2 \to \tilde{M}$ of f and apply 4.11 to obtain an embedding $\tilde{f}_1 : S^2 \to \tilde{M}$ such that $[\tilde{f}_1] \notin p_*^{-1}(N)$ and $\tilde{f}_1(S^2) \subset \tilde{f}(S^2) \cup p^{-1}(U)$. We may assume that $f_1 = p \circ \tilde{f}_1$ is a general position map. Since $\Sigma_0(f_1) = \Sigma_3(f_1) = \emptyset$, every double curve of f_1 is simple. If for some double curve J of f_1 $f_1^{-1}(J)$ has two components, we can apply 4.6 to eliminate J. So we assume that each double curve of f_1 is doubly covered by a curve in $S(f_1)$. Let D be a 2-cell in S^2 such that $D \cap S(f_1) = \partial D$. Now $f_1(D)$ is a projective plane which is 2-sided in M since it contains the orientation reversing curve $f_1(\partial D)$ (see 4.6). Let $g: S^2 \to f_1(D)$ be a double covering map. If $[g] \notin N$ we are done. Otherwise we apply 4.6 (case 2) to obtain a general position map $f_2: S^2 \to M$ such that $f_2(S^2) \subset f_1(S^2) \cup R$ (R a neighborhood of $f_1(\partial D)$) and $c(f_2) <$ (f_1). One checks that $[f_2] \equiv [f_1]$ modulo the $\pi_1(M)$-submodule generated by $[g]$. Thus $[f_2] \notin N$ and the proof follows by induction.

The following theorem can be established by the same sort of technique used above. We refer to [104] for a proof.

4.13. THE GENERALIZED LOOP THEOREM. *Let* M *be a 3-manifold and* D *a compact planar surface with boundary components* $J_1, \cdots, J_k$. *Let* $N_1, \cdots, N_k$ *be normal subgroups of* $\pi_1(M)$. *Suppose* $f: (D, \partial D) \to (M, \partial M)$ *is a map such that* $[f | J_i] \notin N_i$, $f(J_i)$ *is orientation preserving, and* $f(J_i) \cap f(J_j) = \emptyset$ *for* $i \neq j$. *Then given regular neighborhoods* U_i *of* $f(J_i)$ *in* ∂M, *there is a planar surface* E *and an embedding* $g: (E, \partial E) \to (M, \cup U_i)$ *satisfying*

(i) *for no* i *does* U_i *contain the image, under* g, *of more than one component of* ∂E, *and*

(ii) *for some* i *there is a component* L *of* ∂E *such that* $g(L) \subset U_i$ *and* $[g|L] \notin N_i$.

CHAPTER 5
FREE GROUPS

As a first application of the loop and sphere theorems we describe the structure of a compact 3-manifold with free fundamental group. With respect to the projective plane theorem we need

5.1. LEMMA. *If* P *is a projective plane embedded in a* 3-*manifold* M, *then* $i_*: H_1(P; Z_2) \to H_1(M; Z_2)$ *is monic; thus* $i_*: \pi_1(P) \to \pi_1(M)$ *is monic.*

PROOF. Let J be the nontrivial simple loop in P. If P is 2-sided in M, then J is orientation reversing in M. If P is 1-sided in M, then J may be homotoped so as to meet P transversely in a single point. In either case J cannot be homologous to zero (mod 2) in M.

5.2. THEOREM. *Suppose* M *is a prime, compact* 3-*manifold with* $\pi_1(M)$ *a nontrivial free group. Then* M *is either*

(i) *a* 2-*sphere bundle over* S^1, *or*

(ii) *a (possibly nonorientable) cube with handles.*

PROOF. First suppose that $\pi_2(M) \neq 0$. Then by the sphere and projective plane theorems, there is a map $g: S^2 \to \text{Int } M$ such that $g: S^2 \to g(S^2)$ is a covering map and $0 \neq [g] \in \pi_2(M)$. By 5.1, $g(S^2)$ is not a projective plane, since $\pi_1(M)$ is torsion free. Thus g is an embedding. If $g(S^2)$ separates M, we have a corresponding factorization $M = M_1 \# M_2$. One of M_1, M_2 must be S^3, since M is prime. This gives the contradiction that $[g] = 0$. Thus $g(S^2)$ does not separate M. By 3.8, $M = M_1 \# M_2$ where M_1 is a 2-sphere bundle over S^1. Since M is prime, $M = M_1$.

Now suppose $\pi_2(M) = 0$. By applying the Hurewicz theorem to the (noncompact) universal cover of M, we see that $\pi_i(M) = 0$ for $i \geq 2$. Thus M is homotopy equivalent to a 1-dimensional complex and $H_q(M) = 0$ for $q \geq 2$. Thus $\partial M \neq \emptyset$. No component of ∂M is a 2-sphere; otherwise we could factor $M = B^3 \# M_1$ and would have $M = B^3$. Let F be a component of ∂M. Since $\pi_1(F)$ is not free and every subgroup of a free group is free, we have $\ker(\pi_1(F) \to \pi_1(M)) \neq 1$. By the loop theorem there is an embedding $g : (B^2, \partial B^2) \to (M, F)$ such that $g(\partial B^2)$ does not bound a 2-cell in F. Let $M' = M - g(B^2) \times (-1,1)$ be M cut along $g(B^2)$. If M' is connected, then $\pi_1(M')$ is free of rank $r-1$ where $r = \text{rank}\, \pi_1(M)$. If M' has two components M'_1 and M'_2, then $\pi_1(M'_i)$ is free of rank $r_i (i = 1, 2)$ where $r_i > 0$ and $r_1 + r_2 = r$. Observing that the components of M' are prime, the proof that M is a cube with handles is completed by induction on r. One must note that for $r = 1$, $\partial M'$ is connected and in fact $M' = B^3$ (not a fake 3-cell); otherwise we could factor M nontrivially.

Using 5.2 together with the existence and uniqueness of prime factorizations one has, up to connected sum with a homotopy 3-sphere, the structure of an arbitrary compact 3-manifold with free fundamental group. We leave the computation of the number of the various types of prime factors as

5.3. EXERCISE. *Let* M *be a compact* 3*-manifold with* $\pi_1(M)$ *free of rank* $r > 0$. *Let* $F_1, \cdots, F_k$ *be the components of* ∂M. *Show that*

$$M = \Sigma \# C_1 \# \cdots \# C_k \# B_1 \# \cdots \# B_s$$

where C_i *is a cube with* r_i*-handles,* $r_i = \frac{1}{2}(2 - \chi(F_i))$, B_j *is a* 2*-sphere bundle over* S^1, $s = r - \sum_{i=1}^{k} r_i$, *and* Σ *is a homotopy* 3*-sphere.*

Note that the Σ factor may be omitted if M (or any cover of M) contains no fake 3-cell; e.g. if the universal cover of M embeds in R^3.

CHAPTER 6
INCOMPRESSIBLE SURFACES

In Chapter 2 we showed that each closed 3-manifold can be split into two cubes with handles by a 2-manifold (Heegaard surface). The advantage of such a decomposition is obvious – the pieces are simple. However, it is difficult to obtain useful information about the embedding of a Heegaard surface from intrinsic properties of the corresponding manifold. On the other hand, the incompressible surfaces which we consider here are highly representative of the (homotopy theoretic) properties of the corresponding manifolds and play a key role in most subsequent developments. This chapter is devoted to several technical lemmas regarding incompressible surfaces.

By the term *surface* we will mean a compact, connected 2-manifold. Let M be a 3-manifold and F a surface which is either properly embedded in M or contained in ∂M. We say that F *is incompressible in* M if none of the following conditions is satisfied.

(i) F is a 2-sphere which bounds a homotopy 3-cell in M, or

(ii) F is a 2-cell and either $F \subset \partial M$ or there is a homotopy 3-cell $X \subset M$ with $\partial X \subset F \cup \partial M$, or

(iii) there is a 2-cell $D \subset M$ with $D \cap F = \partial D$ and with ∂D not contractible in F.

6.1. LEMMA. *Let* S *be a compact* 2*-manifold in a* 3*-manifold* M *such that each component of* S *is either properly embedded and* 2*-sided in* M *or is contained in* ∂M. *If for some component* F *of* S $\ker(\pi_1(F) \to \pi_1(M)) \neq 1$, *then there is a* 2*-cell* D *in* M *such that* $D \cap S = \partial D$ *and* ∂D *is not contractible in* S.

PROOF. Choose a general position map $f:(B^2, \partial B^2) \to (M, F)$ such that $f|\partial B^2$ is essential (not nullhomotopic) in F. Using the product structure on a neighborhood of S we may assume that for some neighborhood N of ∂B^2 in B^2 $f(N - \partial B^2) \cap S = \emptyset$. Thus the components of $f^{-1}(S)$ are simple closed curves. We suppose that f is chosen to minimize the number of these components. Let E be a 2-cell in B^2 such that $E \cap f^{-1}(S) = \partial E$ and let F' be the component of S containing $f(\partial E)$. It follows that $f|\partial E$ is essential in F'; otherwise we could modify f in a neighborhood of E to contradict our minimality assumption. The conclusion now follows by applying the loop theorem to the manifold obtained by cutting M along F'.

6.2. COROLLARY. *If F is a 2-sided incompressible surface in a 3-manifold M, then* $\ker(\pi_1(F) \to \pi_1(M)) = 1$.

The hypothesis that F be 2-sided is necessary for 6.2. Examples are provided by

6.3. LEMMA. *Let M be an orientable 3-manifold which contains a closed surface with odd Euler characteristic and let F be a closed surface in M with maximal odd Euler characteristic. Then F is incompressible in M. If $F = P^2$, then $\pi_1(M) \cong Z_2 * G$ (where possibly $G = 1$).*

PROOF. If F were compressible, then by doing surgery on a compressing 2-cell D we would obtain either one new surface F' with $\chi(F') = \chi(F) + 2$ or two new surfaces F', F'' with $\chi(F') + \chi(F'') = \chi(F) + 2$ according as ∂D does not or does separate F. Either case would give a contradiction to the choice of F. Note that in any case F is 1-sided in M; since M is orientable whereas F is not. So if $F = P^2$, then the boundary of a regular neighborhood of F is a 2-sphere. This establishes the final claim.

In order to apply 6.3 it suffices to exhibit 3-manifolds with finite fundamental group ($\neq Z_2$) which contain closed surfaces with odd Euler characteristic. We leave this as

6.4. EXERCISE. *Show that the lens space* $L_{2k,q}$ *contains a closed surface* F *with* $\chi(F) = 2-k$.

For a 3-manifold M and a space X we say that two maps $f, g : M \to X$ are C-*equivalent* if there are maps $f = f_0, f_1, \cdots, f_n = g$ of M to X with either f_i homotopic to f_{i-1} or f_i agreeing with f_{i-1} on $M-B$ for some homotopy 3-cell $B \subset M$ with $B \cap \partial M$ empty or a 2-cell. If $\pi_3(X)=0$ C-equivalent maps are homotopic. In any case C-equivalent maps induce the same homomorphism $\pi_1(M) \to \pi_1(X)$ up to choices of base points and inner automorphisms.

6.5. LEMMA. *Suppose* M *is a compact* 3-*manifold and* X *is a p.l.* k-*manifold containing a properly embedded, 2-sided p.l.* (k–1)-*submanifold* Y *with* $\ker(\pi_1(Y) \to \pi_1(X)) = 1$ *and* $\pi_2(Y) = \pi_2(X-Y) = 0$. *If* $f : M \to X$ *is any map, then there is a map* $g : M \to X$ *such that*

(i) g *is C-equivalent to* f,

(ii) *each component of* $g^{-1}(Y)$ *is a properly embedded, 2-sided, incompressible surface in* M, *and*

(iii) *for properly chosen product neighborhoods* $Y \times [-1,1]$ *of* $Y = Y \times 0$ *in* X *and* $g^{-1}(Y) \times [-1,1]$ *of* $g^{-1}(Y) = g^{-1}(Y) \times 0$ *in* M, g *maps each fiber* $p \times [-1,1]$ *homeomorphically to the fiber* $g(p) \times [-1,1]$ *for each* $p \in g^{-1}(Y)$.

PROOF. We suppose that X is triangulated in such a way that a product neighborhood $Y \times [-1,1]$ of Y is triangulated as a product – each simplex being the join of a simplex in $Y \times (-1)$ with a simplex in $Y \times 1$. Then $Y = Y \times 0$ is "transverse" with respect to this triangulation. After adjusting by a homotopy, we may assume that f is simplicial as a map from some triangulation of M to this triangulation of X. One checks that each component of $f^{-1}(Y)$ is a properly embedded 2-sided surface and that condition (iii) is satisfied relative to $Y \times [-\frac{1}{2}, \frac{1}{2}]$ and $f^{-1}(Y \times [-\frac{1}{2}, \frac{1}{2}])$. If each component of $f^{-1}(Y)$ is incompressible, we are done. If not, we consider the following three possibilities.

(1) $f^{-1}(Y)$ contains a compressible 2-sphere F. Then F bounds a homotopy 3-cell C in M. Since $\pi_2(Y) = 0$ we may construct a map $f_1 : M \to X$ such that $f_1 | M - U = f | M - U$ (U a small regular neighborhood of C) and $f_1(U) \subset Y \times [-1,1] - Y$. Then f_1 is C-equivalent to f and $f_1^{-1}(Y) \subset f^{-1}(Y) - F$.

(2) $f^{-1}(Y)$ contains a compressible 2-cell F. As in case (1) we obtain a map $f_1 : M \to X$ C-equivalent to f with $f_1^{-1}(Y) \subset f^{-1}(Y) - F$.

(3) There is a 2-cell D in Int M with $D \cap f^{-1}(Y) = \partial D$ and ∂D not contractible in $f^{-1}(Y)$. We choose a regular neighborhood C of D in M such that $A = C \cap f^{-1}(Y)$ is an annulus properly embedded in C. Let D_1 and D_2 be the disjoint 2-cells in ∂C with $\partial A = \partial D_1 \cup \partial D_2$, and choose disjoint 2-cells E_1 and E_2 properly embedded in C with $\partial E_i = \partial D_i$. We define $f_1 : M \to X$ as follows. Put $f_1 | M - \text{Int } C = f | M - \text{Int } C$. Since $\ker(\pi_1(Y) \to \pi_1(X)) = 1$, we may extend $f_1 | \partial E_i$ to map E_i into Y. Using the product structure on a neighborhood of Y and the assumption that $\pi_2(X - Y) = 0$, we may extend f_1 over each of the components of $C - E_1 \cup E_2$ in such a way that $f_1^{-1}(Y) \cap C = E_1 \cup E_2$. Then f_1 is C-equivalent to f and $f_1^{-1}(Y) = (f^{-1}(Y) - A) \cup E_1 \cup E_2$. Note that f_1 is forced to map opposite sides of E_i to opposite side of Y.

In each of the above cases we may use regular neighborhood theory to homotope f_1 (without changing $f_1^{-1}(Y)$) to a map which satisfies condition (iii) of the theorem. Thus the proof will be completed by an inductive argument provided that we show that f_1 is "simpler" than f. As one of the many possible ways of measuring the complexity of f we use

$$c(f) = (\cdots, n_{-1}, n_0, n_1, n_2)$$

where n_i is the number of components of $f^{-1}(Y)$ having Euler characteristic i. We order complexities lexicographically: i.e.,

$$(\cdots, m_{-1}, m_0, m_1, m_2) < (\cdots, n_{-1}, n_0, n_1, n_2)$$

if there is some k such that

$$m_i = n_i \quad \text{for} \quad i < k \quad \text{and} \quad m_k < n_k \ .$$

We leave it to the reader to verify that in every case $c(f_1) < c(f)$ to finish the proof.

Observe that if a 3-manifold M contains a properly embedded, 2-sided, nonseparating surface F, then $H_1(M)$ is infinite (any closed curve meeting F transversely in a single point represents an element of infinite order in $H_1(M)$). The next two lemmas give a strong converse to this result.

6.6. LEMMA. *Suppose that* M *is a compact 3-manifold and that* $H_1(M)$ *is infinite. Then* M *contains a properly embedded, 2-sided, nonseparating incompressible surface.*

PROOF. By assumption there is an epimorphism of $\pi_1(M)$ onto Z and hence a map $f: M \to S^1$ which induces an epimorphism $f_*: \pi_1(M) \to \pi_1(S^1)$. By 6.5 we may assume that for some point $y \in S^1$ each component of $f^{-1}(y)$ is a properly embedded, 2-sided, incompressible surface in M. Note that $f^{-1}(y) \neq \emptyset$, since f_* is epic. Using (iii) of 6.5 we can choose a generator z for $\pi_1(S^1)$ such that for any loop α in M which meets $f^{-1}(y)$ transversely $f_*([\alpha]) = z^n$ where n is the sum of the "signed" intersection numbers of α with $f^{-1}(y)$. If we choose α so that $f_*([\alpha]) = z$, we see that α must cross some component F of $f^{-1}(y)$ an odd number of times. Thus $M - F$ is connected and the proof is complete.

In many cases the hypothesis that $H_1(M)$ be infinite is automatically satisfied. Specifically we have

6.7. LEMMA. *Let* M *be a compact* 3-*manifold and suppose either*

(i) M *is orientable and* $\partial\hat{M} \neq \emptyset$, *or*

(ii) M *is nonorientable and* $\partial\hat{M}$ *contains no projective plane (possibly* $\partial\hat{M} = \emptyset$).

Then $H_1(M)$ *is infinite.*

PROOF. In every case $\beta_3(\hat{M}) = 0$, and $\chi(\hat{M}) = \frac{1}{2}\chi(\partial\hat{M}) \leq 0$. Thus $\beta_1(M) = \beta_1(\hat{M}) \geq 1 + \beta_2(\hat{M}) \geq 1$.

In the bounded, orientable case we have a stronger version of 6.6 given by

6.8. LEMMA. *Suppose* M *is a compact, orientable* 3-*manifold and that* ∂M *contains a surface of positive genus. Then* M *contains a properly embedded,* 2-*sided, incompressible surface* F *such that* $0 \neq [\partial F] \in H_1(\partial M)$ *(hence* $\partial F \neq \emptyset$).

PROOF. We first observe that since M is orientable and F is (to be) 2-sided, F will be orientable. Thus an orientation on F will induce one on ∂F; so we have a well determined, up to sign, element $[\partial F] \in H_1(\partial M)$.

We proceed as in 6.6: Choose a map $f: M \to S^1$ such that for some $y \in S^1$ each component of $f^{-1}(y)$ is incompressible. Suppose that for some component F of $f^{-1}(y)$ the conclusion fails for F. Then ($\partial F = \emptyset$, or) ∂F is the *oriented* boundary of some submanifold of ∂M. Using the fact that M is orientable we see that for any loop α in ∂M the sum of the oriented intersection numbers of α with ∂F is zero. Thus if the conclusion fails for every component of $f^{-1}(y)$, we must have $f_*(H_1(\partial M)) = 0 \in H_1(S^1) = \pi_1(S^1)$. Thus the proof will follow once we show how to choose f so that $f_*(H_1(\partial M)) \neq 0$. For this it suffices to show that image $(i_*: H_1(\partial M) \to H_1(M))$ is infinite.

If this is not the case, then image $(i_*: H_1(\partial M; Q) \to H_1(M; Q)) = 0$; where Q denotes the rationals. Thus we have an exact sequence (with rational coefficients)

$$0 \to H_3(M,\partial M) \to H_2(\partial M) \to H_2(M) \to H_2(M,\partial M) \to H_1(\partial M) \to 0 .$$

The alternating sum of the ranks is zero. Since $H_2(M,\partial M) = H^1(M) = H_1(M)$, by duality, we have

$$1 - p_2(\partial M) + p_2(M) - p_1(M) + p_1(\partial M) = 0$$

where the p_i's are rational betti numbers. Thus $\chi(M) = \chi(\partial M) - p_0(\partial M)$. Since $\chi(M) = \frac{1}{2}\chi(\partial M)$, we have the contradiction that every component of ∂M is a 2-sphere.

Some words of caution regarding the last two lemmas are appropriate. First, 6.8 does not assert that ∂F does not bound a surface in ∂M only that ∂F (as oriented by F) is not the oriented boundary of a surface in ∂M. Second, 6.7 does not apply to nonorientable 3-manifolds whose boundary contains projective planes – even if the boundary also contains orientable surfaces of positive germs. Finally if M is nonorientable and $\partial \hat{M} \neq \emptyset$, one cannot assume that image $(H_1(\partial M) \to H_1(M))$ is infinite even if ∂M contains no projective plane. Examples are given below.

6.9. EXAMPLE. Let M be the twisted I-bundle over the Klein bottle. Then M is orientable. If F is a 2-sided incompressible surface in M with $\partial F \neq \emptyset$, then, since ∂M is incompressible and $\pi_1(M)$ does not contain a free subgroup of rank greater than one, F is an annulus. Thus ∂F bounds annuli B_1, B_2 in $\partial M = S^1 \times S^1$. However $F \cup B_i$ is a Klein bottle $(i=1,2)$.

6.10. EXAMPLE. Let F be a closed orientable surface and $\tau : F \to F$ an involution such that $\tau_* : H_1(F) \to H_1(F)$ sends each element to its inverse and such that τ has a finite number of fixed points $x_1, \cdots, x_n$. For example let $F = S^1 \times S^1$ and $\tau(e^{i\theta}, e^{i\phi}) = (e^{-i\theta}, e^{-i\phi})$. Define $\sigma : F \times [-1,1] \to F \times [-1,1]$ by $\sigma(x,t) = (\tau(x), -t)$. Let $B_1, \cdots, B_n$ be invariant 3-cell neighborhoods of $x_1 \times 0, \cdots, x_n \times 0$. Then σ acts freely on

$F \times [-1,1] - \cup \text{Int } B_i$. Let $M = (F \times [-1,1] - \cup \text{Int } B_i)/\sigma$. Then ∂M contains n projective planes and one copy of F. It is easy to show that $H_1(M)$ is finite.

6.11. EXAMPLE. Let V be a solid Klein bottle and T a solid torus in Int V as shown below.

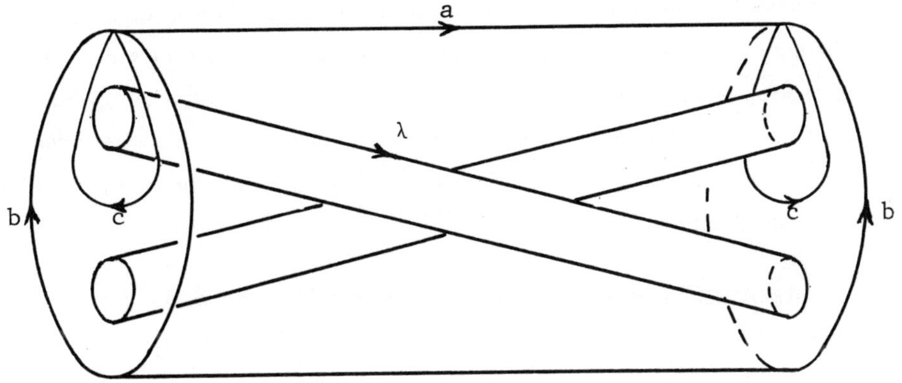

Then $\pi_1(V - \text{Int } T) = \langle a, b, c : aba^{-1} = b^{-1}, aca^{-1} = b^{-1}c \rangle$. Furthermore, a properly chosen longitude λ of ∂T represents a^2. We attach a 2-handle to $V - \text{Int } T$ along λ to obtain M. One checks that image $(H_1(\partial M) \to H_1(M)) = Z_2$.

CHAPTER 7
KNESER'S CONJECTURE ON FREE PRODUCTS

The main goal of this chapter is to prove that each splitting of the fundamental group of a closed 3-manifold as a free product is induced by splitting of the manifold as a connected sum. A proof could be based on the existence and uniqueness of connected sum decompositions together with Grushko's theorem. However, for completeness and to illustrate some useful technique, we give a proof which is based on a topological proof of Grushko's theorem [97].

7.1. THEOREM. *Let* M *be a compact* 3-*manifold such that each component of* ∂M *(possibly empty) is incompressible in* M. *If* $\pi_1(M) \cong G_1 * G_2$, *then* $M = M_1 \# M_2$ *where* $\pi_1(M_i) \cong G_i$, $i = 1, 2$.

PROOF. Choose complexes X_1 and X_2 with $\pi_1(X_i) \cong G_i$ and $\pi_2(X_i) = 0$. Join a point of X_1 to a point of X_2 by a 1-simplex A to form a complex $X = X_1 \cup A \cup X_2$. Note that $\pi_1(X) \cong G_1 * G_2$ and $\pi_2(X) = 0$. Thus we can construct a map $f: M \to X$ such that $f_*: \pi_1(M) \to \pi_1(X)$ is an isomorphism (which can be preassigned). Choose $x_0 \in \text{Int } A$. By 6.5 we may assume that each component of $f^{-1}(x_0)$ is a 2-sided incompressible surface properly embedded in M. If F is a component of $f^{-1}(x_0)$, then since $\ker(\pi_1(F) \to \pi_1(M)) = 1$ (6.2), f_* is monic, and $f(F) = x_0$, we must have $\pi_1(F) = 1$. If some component F of $f^{-1}(x_0)$ is a (incompressible) 2-cell, then by hypothesis ∂F bounds a 2-cell $D \subset \partial M$. The 2-sphere $F \cup D$ can be pushed slightly into Int M to obtain an incompressible 2-sphere F'. Since $\pi_2(X_i) = 0$, f can be modified by a C-equivalence, to a map which replaces F by F' as a component of the inverse of x_0.

By this reasoning we may now assume that each component of $f^{-1}(x_0)$ is an incompressible 2-sphere in Int M. If $f^{-1}(x_0)$ is connected, we are done. If not, there is a path $\beta : I \to M$ such that $\beta(0)$ and $\beta(1)$ lie in different components of $f^{-1}(x_0)$. Now $f \circ \beta$ is a loop in X and since f_* is epic, there is a loop γ based at $\beta(1)$ such that $[f \circ \gamma] = [f \circ \beta]^{-1}$. Then $\alpha = \beta\gamma$ is a path satisfying

(i) $\alpha(0)$ and $\alpha(1)$ are in different components of $f^{-1}(x_0)$,

(ii) $[f \circ \alpha] = 1 \in \pi_1(X)$.

We may assume that α is a simple path which crosses $f^{-1}(x_0)$ transversely at each point of $\alpha(\text{Int } I)$. Of all such paths satisfying the above conditions we assume that $\#(\alpha^{-1}(f^{-1}(x_0)))$ is minimal. We must have $\alpha(\text{Int } I) \cap f^{-1}(x_0) = \emptyset$. For if not, we can write $\alpha = \alpha_1\alpha_2 \cdots \alpha_k$ $(k \geq 2)$ where for each i, $\alpha_i(\text{Int } I) \cap f^{-1}(x_0) = \emptyset$ and $\alpha_i(\partial I) \subset f^{-1}(x_0)$. Then $[f \circ \alpha_1][f \circ \alpha_2] \cdots [f \circ \alpha_k]$ is a representation of the identity element as an alternating product in the free product $G_1 * G_2$. Thus for some i $[f \circ \alpha_i] = 1$. If $\alpha_i(0)$ and $\alpha_i(1)$ lie in the same component of $f^{-1}(x_0)$, we could reduce $\#\alpha^{-1}(f^{-1}(x_0))$. If not, we contradict our minimality assumption. Thus we have $\alpha(\text{Int } I) \cap f^{-1}(x_0) = \emptyset$. Let $F_j (j = 0,1)$ be the component of $f^{-1}(x_0)$ containing $\alpha(j)$. Let C be a small regular neighborhood of $\alpha(I)$ such that $C \cap F_j = D_j$ is a spanning 2-cell of C and $C \cap f^{-1}(x_0) = D_0 \cup D_1$. Let B be the annulus in ∂C bounded by $\partial D_0 \cup \partial D_1$. Push Int B slightly into Int C to obtain an annulus B' with $\partial B' = \partial B$ and $B \cup B'$ the boundary of a solid torus T. We define a map $f_1 : M \to X$ as follows. Put $f_1 | M - \text{Int } C = f | M - \text{Int } C$ and $f_1(B') = x_0$. Since $[f \circ \alpha] = 1$, we can extend f_1 across a meridional 2-cell E of T. Now it remains to extend f_1 across the remaining two open 3-cells; this can be done since $\pi_2(X_i) = 0$ $i = 1, 2$. The extension can be done so that $f_1^{-1}(x_0) \cap C = B'$. Thus f_1 is C-equivalent to f and $f_1^{-1}(x_0) = (f^{-1}(x_0) - (D_0 \cup D_1)) \cup B'$ has one less component than $f^{-1}(x_0)$. The proof is completed by induction.

If we drop the hypothesis of boundary incompressibility from 7.1, the situation does not become impossible, but is more awkward to describe. Note that free products can be introduced by taking disk connected sums, and that a free product with Z can be introduced by adding a 1-handle $(B^2 \times I)$. In the second case, if $B^2 \times 0$ and $B^2 \times 1$ are attached to the same boundary component, we get a disk connected sum; while if they are attached to different boundary components, we do not. We refer to [49] for a discussion of this situation.

Generalizations of Kneser's conjecture to free products amalgamated along the fundamental group of a 2-manifold and to certain HNN groups can be made by varying the above techniques. See [24] and [25] for a treatment of these generalizations.

CHAPTER 8
FINITELY GENERATED SUBGROUPS

Let M be a (possibly noncompact) 3-manifold. Suppose H is a subgroup of $\pi_1(M)$. Clearly H is the fundamental group of some 3-manifold (a covering of M). We consider the question: When is H the fundamental group of some compact 3-manifold? An obvious necessary condition is that H be finitely presented. We present here the striking result that it is sufficient to assume only that H be finitely generated. This should be contrasted with the fact that many finitely presented groups have finitely generated nonfinitely presentable subgroups – including such "simple groups" as $F_2 \times F_2$ (F_2 free of rank 2).

We proceed in two steps. The first, due to Jaco [51], is

8.1. THEOREM. *Let* M *be a (possibly noncompact) 3-manifold and* H *be a finitely presented subgroup of* $\pi_1(M)$. *Then there is a compact 3-manifold* R *and an immersion* $f: R \to M$ *such that* $f_*: \pi_1(R) \to \pi_1(M)$ *is monic with image* H.

PROOF. There is a finite 2-complex K with $\pi_1(K) \cong H$ and a map $g: K \to M$ such that $g_*: \pi_1(K) \to \pi_1(M)$ is monic with image H. Let V_0 be a regular neighborhood of g(K) in M and $g_0 = g: K \to V_0$. If $g_{0*}: \pi_1(K) \to \pi_1(V_0)$ is epic (hence an $\cong$), put $R = V_0$ and $f = i: V_0 \to M$. Otherwise, let $p_1: M_1 \to V_0$ be the covering space such that $p_{1*}\pi_1(M_1) = g_{0*}\pi_1(K)$. There is a lifting $g_1: K \to M_1$. Let V_1 be a regular neighborhood of $g_1(K)$ in M_1. We continue, as in the proof of the loop theorem, to construct a tower:

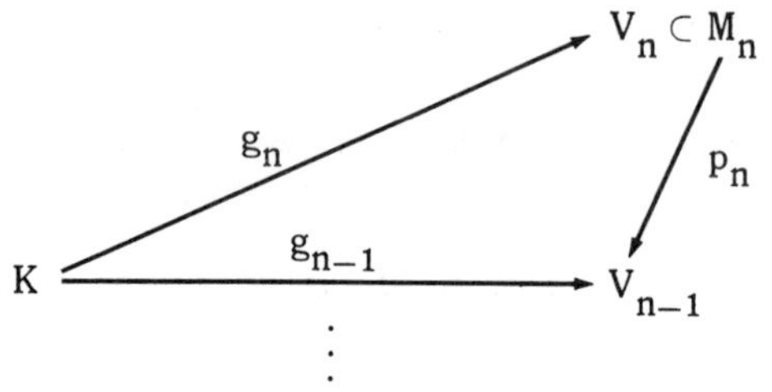

The proof, by comparing the singularities of the maps g_i, given in the loop theorem applies here to show that this construction must terminate at some finite height n. We must have $g_{n*}: \pi_1(K) \to \pi_1(V_n)$ an isomorphism. We put $R = V_n$ and $f = p_1 \circ \cdots \circ p_n | V_n$.

The next theorem is due to G. P. Scott [89].

8.2. THEOREM. *Let* M *be a (possibly noncompact)* 3*-manifold and* H *be a finitely generated subgroup of* $\pi_1(M)$. *Then* H *is finitely presented.*

We first need some preliminaries. For a finitely generated group H we denote by p(H) the minimal number of generators for H. Recall that the equation $p(H_1 * H_2) = p(H_1) + p(H_2)$ is a consequence of Grushko's theorem. From this it follows that any finitely generated group H can be written $H = H_1 * \cdots * H_n$ where each H_i is *indecomposable* (cannot be further factored via free products). The H_i are called *indecomposable factors* of H.

8.3. THE KUROSH SUBGROUP THEOREM. *If* B *is a subgroup of a free product* $A_1 * A_2$, *then* B *is a free product* $B = B_0 * B_1 * B_2 * \cdots$ *of subgroups of* $A_1 * A_2$ *where* B_0 *is free and for* $i \geq 1$ B_i *is conjugate to a subgroup of one of* A_1 *or* A_2.

A proof can be found in [63]. One can also construct a proof by considering $A_1 * A_2$ as the fundamental group of a wedge $X_1 \vee X_2$ ($\pi_1(X_i) = A_i$) and analyzing the covering of $X_1 \vee X_2$ corresponding to B.

We leave the following consequence of 8.3 as

8.4. EXERCISE. *The indecomposable factors of a* (finitely generated) *group are unique up to order and isomorphism. The indecomposable factors not isomorphic to* Z *are unique up to conjugacy.*

We put $n(H)$ the number of indecomposable factors of H and $z(H)$ the number of these which are isomorphic to Z. The *complexity* of H, $c(H)$, is the pair $c(H) = (n(H), z(H))$. We compare complexities lexicographically.

8.5. LEMMA. *If* $\phi : G \to H$ *is an epimorphism of finitely generated groups which restricts to a monomorphism on each indecomposable factor of* G *which is not* ∞*-cyclic, then either* ϕ *is an isomorphism or* $c(H) < c(G)$.

PROOF. Let $G = G_1 * \cdots * G_r * \cdots * G_n$ and $H = H_1 * \cdots * H_s * \cdots * H_m$ be indecomposable factorizations where $G_i \cong Z$ $(H_i \cong Z)$ if and only if $r < i \leq n$ $(s < i \leq m)$. Thus $n = n(G)$, $m = n(H)$, $m - r = z(G)$, $m - s = z(H)$. The assumption that $\phi | G_i$ be monic for $i \leq r$ is independent of the factorization since these factors are unique up to conjugacy.

By the Kurosh subgroup theorem, for each $i \leq r$ there is some $j(i) \leq s$ with $\phi(G_i)$ conjugate to a subgroup of $H_{j(i)}$. Let N be the smallest normal subgroup of G containing $G_1 \cup \cdots \cup G_r$ and M be the smallest normal subgroup of H containing $H_{j(1)} \cup \cdots \cup H_{j(r)}$. Then ϕ induces an epimorphism $\phi : G/N \to H/M$. Note that the projections $G \to G/N$ and $H \to H/M$ split; in particular G/N is free of rank $n - r = z(G)$. By Grushko's theorem $z(G) \geq n(H/M) \geq z(H)$. Since $n(H) = n(H/M) +$ the number of distinct integers in $\{j(1), \cdots, j(r)\}$, we have $c(G) \geq c(H)$. Furthermore $c(G) = c(H)$ if and only if $j(1), \cdots, j(r)$ are all distinct and $\bar{\phi}$ is an isomorphism. In this case the Kurosh subgroup theorem again shows that $\phi | G_1 * \cdots * G_r$ is monic with image $H_1 * \cdots * H_s$ and, hence, that ϕ is an isomorphism.

PROOF OF THEOREM 8.2. We proceed by induction on p(H). The case p(H) = 1 is trivial. Thus assume that p(H) = n > 1, and that any subgroup K of the fundamental group of a 3-manifold, with p(K) < n, is finitely presented. If H decomposes as a free product, the conclusion follows by induction applied to the factors. Thus we suppose that H is indecomposable.

We first show that there is a finitely presented group G and an epimorphism $\theta: G \to H$ satisfying

(*) There is no commutative diagram

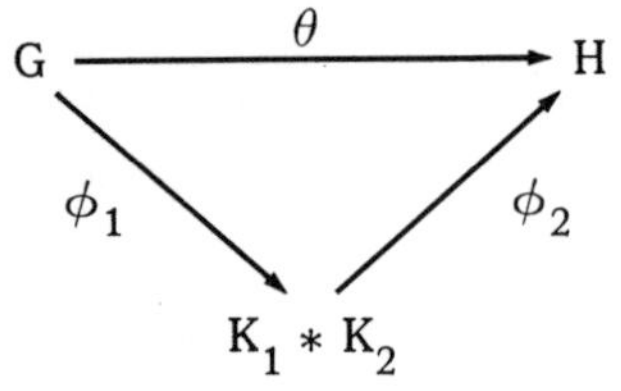

with ϕ_1 an epimorphism and $K_i \neq 1$ $(i = 1,2)$. For this let $\mathcal{C}$ be the collection of all finitely presented groups which admit an epimorphism to H which is monic on each indecomposable factor not isomorphic to Z. So, for example, a free group of rank n is in $\mathcal{C}$. Let G_1 be an element of $\mathcal{C}$ of minimal complexity and $\theta_1 : G_1 \to H$ a corresponding epimorphism. If θ_1 is an isomorphism we are done; otherwise choose $1 \neq g \in \ker \theta_1$, let N be the smallest normal subgroup of G_1 containing g, and let $G = G_1/N$. We have a factorization $\theta_1 = \theta \circ \eta$ where $\eta : G_1 \to G$ is the natural projection. Now $\theta: G \to H$ must satisfy (*); otherwise we have a factorization $\theta = \phi_2 \circ \phi_1$ as in (*). We can naturally factor ϕ_2:

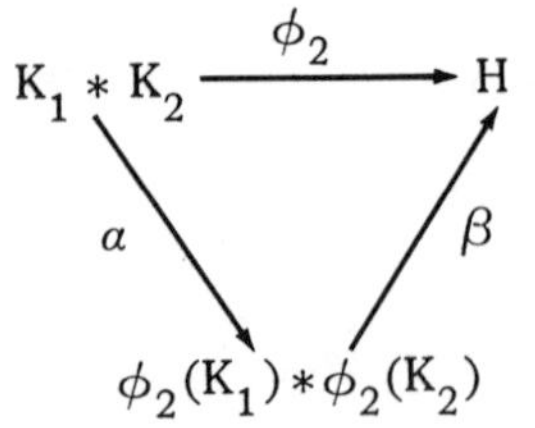

Now $p(\phi_2(K_i)) \leq p(K_i) < n$; so by induction $\phi_2(K_i)$ is finitely presented $(i = 1,2)$. Since $\beta|\phi_2(K_i)$ is monic, and since an indecomposable factorization of $\phi_2(K_1)*\phi_2(K_2)$ can be obtained by factoring $\phi_2(K_1)$ and $\phi_2(K_2)$, it follows that $\phi_2(K_1)*\phi_2(K_2) \in \mathcal{C}$. On the other hand $\phi = \alpha \circ \phi_1 \circ \eta : G_1 \to \phi_2(K_1)*\phi_2(K_2)$ is by construction not monic. Thus by 8.5 $c(\phi_2(K_1)*\phi_2(K_2)) < c(G_1)$. This contradicts the choice of G_1 and completes the proof of the claim.

Now $H \cong \pi_1(R)$ for some 3-manifold R (a covering of M). By the above, there is a finite 2-complex K and a map $f: K \to R$ such that $f_*: \pi_1(K) \to \pi_1(R)$ is an epimorphism satisfying (*). Of all choices for f and all choices of a compact 3-submanifold Q of R containing $f(K)$ we choose a pair (f, Q) for which ∂Q has minimal "complexity." Here we can use the complexity for surfaces used in Lemma 6.5.

We claim that $i_*: \pi_1(Q) \to \pi_1(R)$ is monic. If not, the loop theorem gives a 2-cell D in R with $D \cap \partial Q = \partial D$ and ∂D not contractible in ∂Q. If $\mathrm{Int}\, D \cap Q = \emptyset$, we obtain a "simpler" 3-submanifold Q_1 by adding a regular neighborhood of D to Q. This contradicts our choice of Q. If $D \subset Q$, let Q_1 be a component of Q cut open along D (and Q_2 be the other, if D separates Q). We have $\pi_1(Q) \cong \pi_1(Q_1)*\pi_1(Q_2)$ or $\pi_1(Q) \cong \pi_1(Q_1)*Z$ according as D does or does not separate Q. Since $n > 1$, the splitting is nontrivial. The Kurosh subgroup theorem together with (*) yield that $\mathrm{image}(f_*: \pi_1(K) \to \pi_1(Q))$ is conjugate to a subgroup of $\pi_1(Q_1)$. Thus there is a map $f_1: K \to Q_1$ and an inner automorphism $\gamma: \pi_1(R) \to \pi_1(R)$ such that $f_{1*} = \gamma \circ f_*: \pi_1(K) \to \pi_1(R)$. Thus f_{1*} satisfies (*). Since ∂Q_1 is simpler than ∂Q, we have a contradiction.

Thus $i_*: \pi_1(Q) \to \pi_1(R)$ is monic. Since $f(K) \subset Q$, i_* is also epic. Thus $H \cong \pi_1(Q)$ is finitely presented.

There is a stronger version of 8.2:

8.6. THEOREM. *If* M *is a 3-manifold with* $\pi_1(M)$ *finitely generated, then there is a compact 3-submanifold* Q *of* M *with* $i_*: \pi_1(Q) \to \pi_1(M)$ *an isomorphism.*

Note that the above proof of 8.2 actually establishes 8.6 in the case that $\pi_1(M)$ is indecomposable. The general case is somewhat more involved and we refer to [90] for a proof. We also refer to [28] for conditions that a 3-manifold M contain a compact submanifold Q with $f: Q \to M$ a homotopy equivalence.

CHAPTER 9

MORE ON CONNECTED SUMS; FINITE AND ABELIAN SUBGROUPS

By adding to the results of the last two chapters some elementary facts from the subject of homology of groups, we are able to deduce some theorems on the existence of connected sum decompositions of 3-manifolds which do not involve *a priori* knowledge of a splitting of their fundamental groups. One such result is that the prime factors of an orientable 3-manifold have either finite or torsion-free fundamental groups. We also give a complete list of the abelian groups which occur as fundamental groups of 3-manifolds.

Group Homology

A space X with $\pi_i(X) = 0$ for $i \neq 1$ is *aspherical*. For a group G, an aspherical complex X with $\pi_1(X) \cong G$ is called a K(G, 1). It follows from the Whitehead theorems that, for a fixed group G, any two K(G, 1)'s are homotopy equivalent. Thus we can unambiguously define the i-th *homology group of a group* G, $H_i(G)$, to be $H_i(X)$ where X is any K(G, 1). Of course this agrees with other definitions of the homology of G (with integer coefficients and trivial action) – say via projective resolutions, but we do not need this information. We use only a few elementary facts presented here.

9.1. LEMMA. *If* X *is a connected complex and* $\pi_i(X) = 0$ *for* $1 < i \leq n$, *then* $H_i(X) \cong H_i(\pi_1(X))$ *for* $i \leq n$ *and* $H_{n+1}(\pi_1(X))$ *is a quotient group of* $H_{n+1}(X)$.

PROOF. We can construct a $K(G,1)$ Y from X by successively attaching cells of dimension $\geq n+2$. Then $i_*: H_i(X) \to H_i(Y)$ is an isomorphism for $i \leq n$ and an epimorphism for $i = n+1$.

A complex X as above will be called an *n-dimensional approximation* to a $K(G,1)$ $(G = \pi_1(X))$.

9.2. LEMMA. *For* $i > 0$ *and* $n > 1$

$$H_i(Z_n) = \begin{cases} Z_n; & i \text{ odd} \\ 0\ ; & i \text{ even}. \end{cases}$$

PROOF. There is a free action of Z_n on S^{2k+1} (represent S^{2k+1} as the join of k circles and extend a standard action on each circle). Let $L^{2k+1} = S^{2k+1}/Z_n$; so L^{2k+1} is a 2k-dimensional approximation to a $K(Z_n, 1)$. By duality

$$H_{2k}(L^{2k+1}) \cong H^1(L^{2k+1}) = 0$$

and

$$H_{2k-1}(L^{2k+1}) \cong H^2(L^{2k+1}) \cong H^2(L^3) \cong H_1(L^3) \cong Z_n .$$

9.3. LEMMA. *If* X *is an aspherical complex of dimension* $n < \infty$ *and* G *is any subgroup of* $\pi_1(X)$, *then* $H_i(G) = 0$ *for* $i > n$.

PROOF. Let $\tilde{X}$ be the covering of X corresponding to G. Then $\tilde{X}$ is a $K(G,1)$. Since $\dim \tilde{X} = n$, $H_i(\tilde{X}) = 0$ for $i > n$.

9.4. LEMMA. *If a group* G *has nontrivial torsion, then any* $K(G,1)$ *has infinite dimension.*

PROOF. Apply 9.3 to a finite cyclic subgroup of G and use 9.2.

Finite Groups: The Nonorientable Case

The following striking result was proved by Epstein [20].

9.5. THEOREM. *If* M *is a nonorientable* 3-*manifold with finite fundamental group, then* $\pi_1(M) = Z_2$.

PROOF. We may assume that M is compact, and that ∂M contains no 2-sphere. Let $q: M^* \to M$ be the orientable double cover and $\mu: M^* \to M^*$ the nontrivial covering translation. Duality (cf. 6.7) shows that ∂M^* can contain only 2-spheres. Thus ∂M can contain only projective planes.

(i) ∂M *contains exactly two projective planes*. For, since $\chi(M) = \frac{1}{2}\chi(\partial M)$, ∂M must contain an even number, $2k \geq 0$, of projective planes. Thus ∂M^* contains $2k$ 2-spheres and μ leaves each of these invariant and reverses their orientations. If $k > 0$, $H_2(M^*; Q)$ has a free basis consisting of (any) $2k-1$ of these 2-spheres. Thus the trace of $\mu_*: H_2(M^*; Q) \to H_2(M^*; Q)$ is $-(2k-1)$. Thus we compute the Lefschetz number of μ to be $\lambda(\mu) = 1-0-(2k-1)-0$ if $k > 0$ and $\lambda(\mu) = 1-0+0-(-1) = 2$ if $k = 0$. Since μ is fixed point free, we must have $\lambda(\mu) = 0$. Thus $k = 1$.

(ii) $H_2(\pi_1(M)) = 0$ *and* $H_3(\pi_1(M))$ *is a quotient group of* $Z_2 + Z_2$. Let P_1 and P_2 denote the components of ∂M. Choose two copies P_1^4, P_2^4 of P^4 and identify a standard copy of P^2 in P_i^4 with P_i to obtain a complex $X = M \cup P_1^4 \cup P_2^4$. Now $\pi_1(X) \cong \pi_1(M)$. The universal cover of M is a punctured homotopy 3-sphere. Thus antipodal projections $f_i: S^2 \to P_i$ $(i=1,2)$ serve as a basis for $\pi_2(M)$ as a $\pi_1(M)$-module. Thus $\pi_2(X) = 0$ and 9.1 applies with $n = 2$. The Mayer-Vietoris sequence for $X = M \cup (P_i^4 \cup P_2^4)$ shows that $H_3(X) \cong H_3(P_1^4 \cup P_2^4) \cong Z_2 + Z_2$ and $H_2(X) \cong H_2(M)$. Now $\partial M \neq \emptyset$, so $H_3(M; A) = 0$ for any coefficient group A. Thus, by the universal coefficient theorem, $H_2(M)$ is torsion free. But $\chi(M) = \frac{1}{2}\chi(\partial M) = 1$, and $\beta_1(M) = 0$; so $\beta_2(M) = 0$. Thus $H_2(X) \cong H_2(M) = 0$, and (ii) is established.

Now let $p: \tilde{M} \to M$ be the universal cover and $A (\cong \pi_1(M))$ be the group of covering translations. Let $S_{i1}, \cdots, S_{im}$ be the components of $p^{-1}(P_i)$. Note that $|\pi_1(M)| = 2m$. Now for each (i, j) there is a unique, nontrivial $\sigma_{ij} \in A$ with $\sigma_{ij}(S_{ij}) = S_{ij}$. Since $\tilde{M}/\sigma_{ij}$ is nonorientable, the argument of Step (i) applies to show that there is exactly one other $S_{k\ell}$, say $S_{\mu(i,j),\nu(i,j)}$, left invariant by σ_{ij}.

(iii) *For each* (i, j), $\mu(i, j) \neq i$. For suppose, say, that $\mu(1,1) = 1$. Putting $\nu(1,1) = \nu$, there is some $\tau \in A$ such that $\tau(S_{11}) = S_{1\nu}$. Then $\tau^{-1}\sigma_{11}\tau$ leaves S_{11} invariant. Thus we must have $\tau^{-1}\sigma_{11}\tau = \sigma_{11}$. Let B be the subgroup of A generated by σ_{11} and τ. By the above, B is abelian; so we can write $B = Z_{n_1} + Z_{n_2} + \cdots + Z_{n_r}$ where $n_{i+1} | n_i$. Applying the Künneth formula to a product of $K(Z_{n_i}, 1)$'s and using 9.2, we compute the homology of B. In particular if $r > 1$, $H_2(B)$ contains $Z_{n_2} = H_1(Z_{n_1}) \otimes H_1(Z_{n_2})$ as a summand, and $H_3(B)$ contains Z_{n_1} as a summand. But $M_1 = M/B$ is nonorientable, so (ii) applies to the homology of $\pi_1(M_1) \cong B$. Thus $r = 1$ and $n_1 = 2$; i.e., $B \cong Z_2$. This is a contradiction since $\tau \neq \sigma_{11}$.

(iv) *If* $m > 1$ *(where* $|\pi_1(M)| = 2m$*), then* A *contains a subgroup isomorphic to the dihedral group*

$$D_{2q} = \langle s, t : s^2 = t^q = 1, sts = t^{-1} \rangle$$

for some odd prime q. We first note that for any i, j and any $\tau \in A$, $\tau \neq 1$, σ_{ij}, then $\tau^{-1}\sigma_{ij}\tau \neq \sigma_{ij}$; for suppose, say, that $\tau^{-1}\sigma_{11}\tau = \sigma_{11}$. Now $\tau^{-1}(S_{11}) = S_{1\nu}$ for some $\nu \neq 1$. But then $\sigma_{11} = \tau^{-1}\sigma_{11}\tau$ leaves $S_{1\nu}$ invariant; so we have $\mu(1,1) = 1$ contradicting (iii).

Let $C < A$ correspond to the subgroup of orientation preserving loops in $\pi_1(M)$. Note that C is normal, $\sigma_{ij} \notin C$, and $|C| = m > 1$. If $\tau_1, \tau_2 \in C$ and $\sigma_{11}\tau_1\sigma_{11}\tau_1^{-1} = \sigma_{11}\tau_2\sigma_{11}\tau_2^{-1}$, then $(\tau_2^{-1}\tau_1)^{-1}\sigma_{11}(\tau_2^{-1}\tau_1) = \sigma_{11}$. Since $\sigma_{11} \notin C$, we must have $\tau_2^{-1}\tau_1 = 1$. The elements $\{\sigma_{11}\tau\sigma_{11}\tau^{-1} : \tau \in C\}$ are all distinct and therefore must constitute all of C. From this it follows that for each $\tau \in C$

$$\sigma_{11}\tau\sigma_{11} = \tau^{-1};$$

for by the above, there is some $\xi \in C$ with $\tau = \sigma_{11}\xi\sigma_{11}\xi^{-1}$. Thus $\sigma_{11}\tau\sigma_{11} = \sigma_{11}^2\xi\sigma_{11}\xi^{-1}\sigma_{11} = (\sigma_{11}\xi\sigma_{11}\xi^{-1})^{-1} = \tau^{-1}$ (recall $\sigma_{11}^2 = 1$).

Choose an element τ of prime order q in C. We cannot have $q = 2$; otherwise the equation $\sigma_{11}\tau\sigma_{11} = \tau^{-1} = \tau$ yields the contradiction that $\tau^{-1}\sigma_{11}\tau = \sigma_{11}$. The function $\phi(s) = \sigma_{11}$, $\phi(t) = \tau$ preserves relations and thus extends to a homomorphism $\phi : D_{2q} \to A$. The elements $1, \tau, \cdots, \tau^{q-1}, \sigma_{11}, \sigma_{11}\tau, \cdots, \sigma_{11}\tau^{q-1}$ are all distinct; since $|D_{2q}| = 2q$, ϕ is monic. This establishes (iv).

The proof of Theorem 9.5 is completed by showing that $m = 1$. If $m > 1$, then by (iv) we have $D_{2q} \subset A$ for some odd prime q. Let $M_1 = \tilde{M}/D_{2q}$. Then $\pi_1(M_1) \cong D_{2q}$, and, since $\sigma_{11} \in D_{2q}$, M_1 is nonorientable. Thus (ii) applies to the homology of D_{2q}. We get a contradiction from

(v) $H_3(D_{2q}) = Z_{2q}$. To see this we define an action of D_{2q} on $S^n \times S^5$ by $s(x, z_1, z_2, z_3) = (-x, \bar{z}_1, \bar{z}_2, \bar{z}_3)$ and

$$t(x, z_1, z_2, z_3) = (x, \omega z_1, \omega z_2, \omega z_3)$$

where $x \in S^n$, $(z_1, z_2, z_3) \in S^5 = \{(z_1, z_2, z_3) \in C^3 : |z_1|^2 + |z_2|^2 + |z_3|^2 = 1\}$ and $\omega = e^{\frac{2\pi i}{q}}$. One checks that this action is free and hence $E_n = S^n \times S^5/D_{2q}$ is a closed $(n+5)$-manifold. Note that E_n is orientable precisely when n is even. Now $\pi_1(E_n) = D_{2q}$ for $n > 1$ and $\pi_i(E_4) = 0$ for $1 < i < 4$. Thus $H_3(D_{2q}) \cong H_3(E_4)$. Observe that E_n is a fiber bundle over P^n with fiber a 5-dimensional lens space L^5 and that E_{n-1} is the restriction of this bundle to the standard $P^{n-1} \subset P^n$. Using the homology sequences of the pairs (E_n, E_{n-1}) $(n = 1,2,3,4)$ and noting that $H_q(E_n, E_{n-1}) \cong H_q(L^5 \times B^n, L^5 \times S^{n-1}) \cong H^{n+5-q}(L^5 \times B^n) \cong H_{q-n}(L^5)$, one can compute the homology of E_4 to complete the proof.

9.6. THEOREM. *If M is a compact, nonorientable 3-manifold with finite fundamental group, then M is homotopy equivalent to a punctured $P^2 \times I$ (a manifold of the form $P^2 \times I \# B^3 \# \cdots \# B^3$).*

PROOF. By 9.5, $\pi_1(M) = Z_2$ and ∂M consists of two P^2's together with some S^2's. The inclusion of either P^2 component of ∂M induces a monomorphism, hence isomorphism, of fundamental groups.

By regular neighborhood theory, there is a submanifold N of M which is a punctured $P^2 \times I$ such that $M_1 = \overline{M - N}$ is connected and ∂M_1 consists of two P^2's. The universal cover of M_1 is a homotopy 3-sphere from which 2-open 3-cells have been removed. Thus the inclusion of either component of ∂M_1 into M_1 is a homotopy equivalence. From this it follows that the inclusion $N \to M$ is a homotopy equivalence.

Subgroups with Higher Homology

Let M be a compact, prime 3-manifold. If some component of ∂M is compressible in M, then we may cut M open along a "compressing" 2-cell. By repeating this process, which must terminate, we obtain a collection $D_1, \cdots, D_n$ of pairwise disjoint, properly embedded 2-cells in M such that the components $M_1, \cdots, M_k$ of M cut open along $\cup D_i$ have incompressible boundaries. By Kneser's conjecture (7.1), $\pi_1(M_i)$ is indecomposable with respect to free products. Furthermore each M_i is irreducible unless M is a 2-sphere bundle over S^1. The M_i are called *subprime factors of* M (or of any 3-manifold of which M is a prime factor).

We use $\omega(M)$ to denote the subgroup of $\pi_1(M)$ consisting of all elements represented by orientation preserving loops.

9.7. THEOREM. *Let* M *be a, possibly noncompact, 3-manifold and suppose* G *is a finitely generated subgroup of* $\pi_1(M)$ *satisfying*

(i) $G \not\cong Z$, *and* G *is indecomposable with respect to free products,*

(ii) $H_i(G) \neq 0$ *for some* $i \geq 3$, *and*

(iii) *either* $G < \omega(M)$ *or* G *has no element of order two.*

Then there is a subprime factor R *of* M *whose boundary is either empty or consists entirely of projective planes and with* G *conjugate to a subgroup of finite index in* $\pi_1(R)$.

Furthermore if $|G| = \infty$, *then* $H_3(G) = Z$ *and* $H_i(G) = 0$ *for* $i > 3$.

PROOF. Some compact submanifold of M contains G as a subgroup of its fundamental group; so we assume that M is compact.

Now $\pi_1(M)$ is a free product of a (possibly trivial) free group together with the fundamental groups of its subprime factors. Thus by (i) and the Kurosh subgroup theorem, there is a subprime factor R of M with G conjugate to a subgroup G_1 of $\pi_1(R)$.

Now R is irreducible (recall that R is not embedded in M -- a punctured R is).

Let $p: \tilde{R} \to R$ be the covering with $p_*\pi_1(\tilde{R}) = G_1$. Either $\tilde{R}$ is orientable $(G < \omega(M))$ or $\pi_1(\tilde{R})$ has no element or order two. In either case it follows that $\tilde{R}$ contains no 2-sided projective plane.

Let $\hat{R}$ be the 3-manifold obtained from $\tilde{R}$ by capping off each 2-sphere in $\partial\tilde{R}$ with a 3-cell. Then $\pi_1(\hat{R}) \cong \pi_1(\tilde{R}) \cong G$. By the sphere and projective plane theorems, we have $\pi_2(\hat{R}) = 0$; otherwise there is an essential 2-sphere in $\hat{R}$ contradicting (i). If the universal cover R^* of $\hat{R}$ is closed, then so is $\hat{R}$ and G is finite. If R^* is not closed (e.g., if $|G| = \infty$), then by the Hurewicz theorem, $H_i(R^*) = \pi_i(R^*) = \pi_i(\hat{R}) = 0$ for $i \geq 2$. In this case we have $H_i(G) = H_i(\hat{R})$. But $H_i(\hat{R}) = 0$ for $i > 3$ so we must have, by (ii), $H_3(G) = Z$.

In any case $\hat{R}$ is closed; so $\tilde{R}$ is compact and $\partial\tilde{R}$ contains only 2-spheres. Thus $p: \tilde{R} \to R$ is finite sheeted; so G_1 has finite index in $\pi_1(R)$. Since R is irreducible, ∂R, if not empty, contains only projective planes.

9.8. THEOREM (Epstein [20]). *Let* M *be a possibly noncompact 3-manifold and* G *be a finite subgroup of* $\pi_1(M)$, *then either:*

(i) $G \cong Z_2$ *and there is a 2-sided projective plane* $P \subset M$ *with* G *conjugate to* $i_*\pi_1(P)$, *or*

(ii) $M = M_1 \# R$ *where* R *is closed and orientable,* $\pi_1(R)$ *is finite, and* G *is conjugate to a subgroup of* $\pi_1(R)$.

PROOF. First suppose $G < \omega(M)$. Then G satisfies the hypothesis of 9.7 with the conclusion of 9.4 replacing (ii). The proof of 9.7 works with this change (which we make only to avoid proving that an arbitrary finite group has nonzero homology in infinitely many dimensions). Thus there is a subprime factor R of M with G conjugate to a subgroup, G_1, of finite index in $\pi_1(R)$. Thus $|\pi_1(R)| < \infty$. R must be orientable; otherwise, by 9.5, $\pi_1(R) = Z_2$, so $G_1 = \pi_1(R)$ which contradicts $G < \omega(M)$. Since R is orientable and ∂R can contain only projective planes, $\partial R = \emptyset$. Thus R is in fact a prime factor of M, so we can write $M = M_1 \# R$.

If $G \not< \omega(M)$, then choose $R, \tilde{R}, \hat{R}$ as in the proof of 9.7. Since $\tilde{R}$ is nonorientable, 9.5 applies to show that $G \cong Z_2$. If $\pi_2(\hat{R}) = 0$, the arguments of the previous theorem show that $\tilde{R}$ is compact. Thus $p : \tilde{R} \to R$ is a homeomorphism – otherwise R contradicts 9.5. If $\pi_2(\hat{R}) \neq 0$, then the projective plane theorem gives a 2-sided projective plane $P \subset R$ representing an element of $\pi_2(R) - p_*(\ker(\pi_2(\tilde{R}) \to \pi_2(\hat{R}))$. Since every 2-sphere in $\hat{R}$ is inessential, P must lift homeomorphically to $\tilde{R}$. Now any projective plane in $\tilde{R}$ carries its fundamental group. Thus G is conjugate to $i_*\pi_1(P)$.

9.9. COROLLARY. *Suppose* M *is a prime* 3*-manifold with* $\pi_1(M)$ *infinite. Then any element of finite order in* $\pi_1(M)$ *has order* 2. *If* M *contains no* 2*-sided projective plane, then* $\pi_1(M)$ *is torsion free.*

As a variation of 9.7, we have

9.10. THEOREM. *Let* M *be a* 3*-manifold and* G *a finitely generated subgroup of* $\pi_1(M)$ *satisfying*

(i) G *is indecomposable with respect to free products, and*

(ii) $H_2(G)$ *has nontrivial torsion.*

Then $H_2(G) = Z_2$ *and there is a nonorientable subprime factor* R *of* M *whose boundary consists of at most projective planes and with* G *conjugate to a subgroup of finite index in* $\pi_1(R)$.

PROOF. Clearly $G \ncong Z$, since $H_2(Z) = 0$. Thus we can proceed, as in the proof of 9.7, to construct: a subprime factor R with G conjugate to a subgroup G_1 of $\pi_1(R)$, the covering $p : \tilde{R} \to R$ corresponding to G_1 and the manifold $\hat{R}$ obtained by capping off the 2-sphere components of $\partial\tilde{R}$ with 3-cells. We cannot assume that $\pi_2(\hat{R}) = 0$; however every 2-sphere in $\hat{R}$ is inessential. So by the projective plane theorem, there exist pairwise disjoint 2-sided P^2's, $P_1, \cdots, P_k$ in $\hat{R}$ which generate $\pi_2(\hat{R})$ as a $\pi_1(\hat{R})$-module. We obtain a space X by attaching to $\hat{R}$ copies of P^3 -- identifying $P^2 \subset P^3$ with $P_i (i = 1, \cdots, k)$. The inclusion $\hat{R} \to X$ induces an isomorphism $\pi_1(\hat{R}) \to \pi_1(X)$, and, by Mayer-Vietoris, an isomorphism $H_2(\hat{R}) \to H_2(X)$. Now $\pi_2(X) = 0$; so $H_2(G) \cong H_2(X) \cong H_2(\hat{R})$. But then the universal coefficient theorem and standard arguments about the top dimensional homology of a manifold yield that $\hat{R}$ is closed, nonorientable, and $H_2(\hat{R}) = Z_2$. Thus $\tilde{R}$ is compact, nonorientable and $\partial\tilde{R}$ consists of at most 2-spheres and the conclusion follows as in 9.7.

9.11. THEOREM. *Let* M *be a 3-manifold and* G *a subgroup of* $\pi_1(M)$ *which is isomorphic to the fundamental group of some closed, aspherical 3-manifold. Then there is a subprime factor* R *of* M *whose boundary consists of at most projective planes and with* G *conjugate to a subgroup of finite index in* $\pi_1(R)$.

PROOF. $G \cong \pi_1(Q)$ for some closed aspherical 3-manifold Q. In particular, $H_i(G) \cong H_i(Q)$ for all i. By Kneser's conjecture, G is not a free product, and $G \ncong Z$, since $H_i(Z) = 0$ for $i > 1$.

If Q is orientable, it follows that $G < \omega(M)$ and the conclusion follows from 9.7 since $H_3(G) \cong H_3(Q) \cong Z$.

If Q is nonorientable, then $H_2(G) \cong H_2(Q) \cong Z_2$, and the conclusion follows from 9.10.

9.12. THEOREM (Epstein [20]). *If* $G \cong Z + Z_2$ *is a subgroup of the fundamental group of a 3-manifold* M, *then* $M = M_1 \# R$ *where* R *is closed and nonorientable,* $\pi_1(R) \cong Z + Z_2$, *and* G *is conjugate to a subgroup of* R.

PROOF. Since $H_2(Z+Z_2) \cong H_2(S^1 \times P^3) \cong H_1(S^1) \otimes H_1(P^3) = Z_2$, it follows from 9.10 that G is conjugate to a subgroup G_1 of finite index in $\pi_1(R)$ for some subprime factor R of M whose boundary consists of at most projective planes. Following the proof and using the notation of 9.10, we see that there is a 2-sided projective plane $P \subset \text{Int } R$ which does not represent an element in the $\pi_1(R)$-submodule of $\pi_2(R)$ generated by the components of ∂R; otherwise we would have $\pi_2(\hat{R}) = 0$. It would follow that $\hat{R}$ is aspherical giving the contradiction that $H_i(Z+Z_2) = 0$ for $i > 3$. Now each component of $p^{-1}(P)$ is a projective plane, since each 2-sphere in $\tilde{R}$ is contractible in $\hat{R}$. Let $\tilde{P}$ be a component of $p^{-1}(P)$. Then $\tilde{R} - \tilde{P}$ is connected; otherwise we obtain a compact 3-manifold whose boundary has odd Euler characteristic. It follows that $\pi_1(\tilde{R}-\tilde{P}) = Z_2$; since $\pi_1(\tilde{R})/\pi_1(\tilde{R}-\tilde{P}) = Z$. If C is a component of $R-P$ and $\tilde{C}$ a component of $p^{-1}(C)$, then $i_*: \pi_1(\tilde{C}) \to \pi_1(\tilde{R}-\tilde{P})$ is monic. Thus $\pi_1(C)$ is finite. By 9.5, $\pi_1(C) \cong Z_2$. If $R - P$ has two components, we get the contradiction that $\pi_1(R) = Z_2$. Since $\partial(R-P\times(-1,1))$ contains exactly two projective planes, $\partial R = \emptyset$. By Van Kampen's theorem, $\pi_1(R) = Z+Z_2$.

Abelian Groups

9.13. THEOREM. *Let* G *be a finitely generated abelian group. If* G *is a subgroup of* $\pi_1(M)$ *for some 3-manifold* M, *then* G *is isomorphic to one of*: Z, $Z+Z$, $Z+Z+Z$, Z_p, $Z+Z_2$.

PROOF. For any subgroup $A < G$ there is compact 3-manifold M_A with $\pi_1(M_A) \cong A$. Furthermore $\pi_2(M_A) = 0$, unless A has 2-torsion. The proof is completed by observing that any finitely generated abelian group not listed in the conclusion contains a subgroup prohibited by one of the following cases.

(i) $A = Z+Z+Z+Z \not< G$; otherwise M_A would be aspherical (since A is infinite) and $H_i(A) = 0$ for $i > 3$. However, $H_4(A) = H_4(S^1 \times S^1 \times S^1 \times S^1) = Z$.

(ii) $A = Z+Z_p \not< G$ for $p > 2$; otherwise we could, by 9.8 write $A \cong \pi_1(M_A)$ as a nontrivial free product.

(iii) $A = Z+Z+Z_2 \not< G$; otherwise by 9.12 we could write A as a nontrivial free product.

(iv) $A = Z_r + Z_s \not< G$ where $r | s$; for $H_2(Z_r + Z_s) \cong H_1(Z_r) \otimes H_1(Z_s) \cong Z_r$, so by 9.10, M_A is nonorientable. This contradicts 9.5, since $A \not\cong Z_2$.

We conclude this chapter by establishing one of the few results which can be obtained about nonfinitely generated groups in this context.

9.14. THEOREM (Evans and Moser [23]). *If* G *is a nonfinitely generated abelian group which is a subgroup of* $\pi_1(M)$ *for some 3-manifold* M, *then* G *is isomorphic to a subgroup of the additive group of rational numbers.*

PROOF. We will show that each finitely generated subgroup of G is infinite cyclic. Then, since G is countable, we can write $G = G_1 \cup G_2 \cup \cdots$ where each G_i is infinite cyclic and $G_i \subset G_{i+1}$. Choose a generator z_i for G_i; then $z_i = z_{i+1}^{n_i}$ for some n_i. We get the embedding $\phi : G \to Q$ by $\phi(z_1) = 1$, $\phi(z_{i+1}) = \frac{1}{n_1 n_2 \cdots n_i}$.

To proceed, we assume that $G = \pi_1(M)$ where M is a noncompact 3-manifold.

(i) G *is torsion free.* Otherwise G contains a subgroup A which is either finite, abelian, and of order greater than two or is $Z + Z_2$. Applying 9.8 in the first case or 9.12 in the second, we can write $M = M_1 \# R$ where A has finite index in $\pi_1(R)$. Since $\pi_1(R)$ is finitely generated and $\pi_1(M)$ is not, $\pi_1(M_1) \neq 1$. This gives a contradiction, since an abelian group cannot be a nontrivial free product.

(ii) $Z+Z+Z+$ *is not a subgroup of* G; *otherwise, since* $Z+Z+Z$ is the fundamental group of a closed aspherical 3-manifold $(S^1 \times S^1 \times S^1)$,

9.11 applies. By (i) M can contain no projective plane; so we have the contradiction $M = M_1 \# R$ as in (i).

(iii) $Z+Z$ *is not a subgroup of* G; for suppose the contrary. Then there is a covering $p: \tilde{M} \to M$ with $\pi_1(\tilde{M}) \cong Z+Z$. Let A be the group of covering translations of $\tilde{M}$. If A contains an element of infinite order, then, since $A \cong \pi_1(M)/p_*\pi_1(\tilde{M})$, it follows that $Z+Z+Z$ is a subgroup of G contradicting (ii). We proceed to show that A has an element of infinite order.

By the proof of 8.2, there is a compact 3-manifold $N \subset \operatorname{Int} \tilde{M}$ such that $i_*: \pi_1(N) \to \pi_1(\tilde{M})$ is an isomorphism and each component of ∂N is either a 2-sphere or is incompressible in $\tilde{M}$. If F is a component of ∂N, then $\pi_1(F)$ is a subgroup of $Z+Z$; thus F is a torus or a 2-sphere. Let $\hat{N}$ be obtained from N by capping off the 2-spheres in ∂N with 3-cells. By the sphere and projective plane theorems $\pi_2(\hat{N}) = 0$. Since $\pi_1(\hat{N})$ is infinite, $\hat{N}$ is aspherical. So $H_i(\hat{N}) = H_i(Z+Z)$; so $H_2(\hat{N}) = Z$, $H_3(\hat{N}) = 0$. Thus $\hat{N}$ is orientable and has nonempty boundary. Let F (a torus) be a component of $\partial\hat{N}$. We must have $\pi_1(F) \to \pi_1(N)$ epic; otherwise by attaching a 2-handle to N, we obtain an orientable 3-manifold with fundamental group $Z+Z_n$ $(n > 1)$. By 9.12 or 9.13, this is impossible. So $\pi_1(F) \to \pi_1(\tilde{M})$ is an isomorphism. From this it follows that F separates $\tilde{M}$. Now A is infinite and only a finite number of translates of any compact subset of $\tilde{M}$ can intersect the set. So we may choose pairwise disjoint translates F_0, F_1, F_2 of F in $\tilde{M}$. We claim that some one of these must separate the other two in $\tilde{M}$. If not, then there is a connected 3-manifold $R \subset \tilde{M}$ with $F_i \subset \partial R$ and $\pi_1(F_i) \to \pi_1(R)$ an isomorphism for $i = 0, 1, 2$. Let α, β be a pair of simple loops in F_0 which meet transversely in a single point. α is freely homotopic to a loop in F_1. Thus β has intersection number 1 with the relative 2-cycle determined by this homotopy. However β is freely homotopic to a loop in F_2. This contradicts the invariance of intersection numbers.

Thus we assume that F_0 separates F_1 from F_2 in $\tilde{M}$. Let U_1 and U_2 be the components of $\tilde{M} - F_0$. We assume that $F_i \subset U_i$ $i = 1, 2$.

Choose $\tau_i \in A$ such that $\tau_i(F_0) = F_i$. If $U_1 \supset \tau_1(U_1)$, then by iterating: $U_1 \supsetneqq \tau_1(U_1) \supsetneqq \tau_1^2(U_1) \supsetneqq \cdots$; so τ_1 has infinite order. Similarly if $U_2 \supset \tau_2(U_2)$, τ_2 has infinite order. So assume that $U_1 \subset \tau_2(U_2)$ and $U_2 \subset \tau_1(U_1)$. Since $\tau_1(U_2)$ contains points of U_1 and is disjoint from F_0, we have $\tau_1(U_2) \subset U_1$. Thus $\tau_2^{-1}\tau_1(U_2) \subset \tau_2^{-1}(U_1) \subset \tau_2^{-1}\tau_2(U_2) = U_2$, and $\tau_2^{-1}\tau_1$ has infinite order.

This completes the proof of (iii) and of the theorem; since (i), (ii), and (iii) together show that every finitely generated subgroup of G is infinite cyclic.

9.15. EXERCISE. *If* G *is a subgroup of* Q, *construct a* 3-*manifold* M *with* $\pi_1(M) \cong G$. *(One can take* M *to be the complement, in* S^3, *of a solenoid: the intersection of a properly chosen, decreasing sequence of solid tori).*

CHAPTER 10
I-BUNDLES

In this chapter we prove that a compact 3-manifold whose fundamental group is isomorphic to the fundamental group of a surface ($\neq P^2$) is essentially (explained below) a fiber bundle over a surface with fiber the unit interval I. We extend this result to 3-manifolds whose fundamental group contains the fundamental group of a surface as a subgroup of finite index.

There are some obvious ways of changing a 3-manifold without changing its fundamental group: removing (or adding) a 3-cell, or taking connected sum with a homotopy 3-sphere. However, to each compact 3-manifold there is a unique 3-manifold $\mathcal{P}(M)$ satisfying

(i) $\mathcal{P}(M)$ contains no fake 3-cell and

(ii) $\hat{M} = \mathcal{P}(M) \# \Sigma$ where $\hat{M}$ is obtained from M by capping off the 2-spheres in ∂M with 3-cells and Σ is a homotopy 3-sphere. We call $\mathcal{P}(M)$ the *Poincaré associate* of M. We always assume that M minus a homotopy 3-cell is embedded in $\mathcal{P}(M)$.

A 3-manifold M is said to be P^2-*irreducible* if M is irreducible (every 2-sphere in M bounds a 3-cell in M) and M contains no 2-sided projective plane.

10.1. LEMMA. *Suppose* M *is a compact* 3-*manifold satisfying*

(i) $\pi_1(M) \not\cong Z$ *and* $\pi_1(M)$ *is indecomposable with respect to free products, and*

(ii) *either* M *is orientable or* $\pi_1(M)$ *contains no element of order two.* *Then* $\mathcal{P}(M)$ *is* P^2-*irreducible.*

PROOF. $\hat{M}$ can have only one nonsimply connected prime factor. The other factors are all closed and can be lumped together to form a homotopy 3-sphere. Since $\pi_1(M) \not\cong Z$, the nontrivial factor is irreducible (3.13). By (ii), this factor can contain no 2-sided projective plane.

Products

10.2. THEOREM. *Let* M *be a compact* 3-*manifold and* F *a compact, connected* 2-*manifold* $(\neq B^2, S^2, P^2)$ *in* ∂M *such that* $i_*: \pi_1(F) \to \pi_1(M)$ *is an isomorphism. Then* $\mathcal{P}(M)$ *is homeomorphic to* $F \times I$ *by a homeomorphism which takes* F *to* $F \times 0$.

PROOF. Note that the exclusions $F \neq B^2, S^2$ are trivial; for then $\pi_1(M) = 1$ and $\mathcal{P}(M) = S^3$. However, our proof breaks down for the case $F = P^2$. The Poincaré conjecture is involved here: if there is a fake 3-sphere Σ and an involution $\tau: \Sigma \to \Sigma$ with exactly two fixed points, then $\mathcal{P}((\Sigma - \text{Int}(B_1 \cup B_2))/\tau) \neq P^2 \times I$ $(B_1, B_2$ small, invariant neighborhoods of the fixed points).

For convenience we assume $M = \mathcal{P}(M)$. Then M must be P^2-irreducible; otherwise M would contain an essential 2-sphere and we could not have $i_*: \pi_1(F) \to \pi_1(M)$ epic.

Observe that (each component of) $\partial M - F$ is incompressible in M; otherwise we could express $\pi_1(M)$ as a free product with $i_*\pi_1(F)$ contained in one factor.

Since $i_*: \pi_1(F) \to \pi_1(M)$ is an isomorphism and $\pi_2(F) = 0$, there is a retraction $\rho: M \to F$; note that $\rho_*: \pi_1(M) \to \pi_1(F)$ is an isomorphism.

Case 1: $\partial F \neq \emptyset$. The important thing to prove in this case is that F can be fixed as one end of the product. For it follows readily that M is a cube with handles and M is homeomorphic to $F_1 \times I$ for any bounded surface F_1 with $\chi(F_1) = \chi(F)$ and with F_1 orientable or not according as F is orientable or not.

Choose pairwise disjoint, properly embedded arcs $A_1, \cdots, A_k$ in F which cut F into a 2-cell. By 6.5 we may assume that each component of $\rho^{-1}(\cup A_i)$ is a properly embedded, 2-sided incompressible surface in M. Since ρ_* is monic, each component of $\rho^{-1}(\cup A_i)$ is simply connected. Let D_i be the component of $\rho^{-1}(A_i)$ containing A_i (actually $\rho^{-1}(A_i)$ is connected, but we do not need this). Then D_i is a 2-cell. Choose a product neighborhood $\cup D_i \times I$ of $\cup D_i$ with $\cup \partial D_i \times I \subset \partial M$ and $\cup(D_i \times I) \cap F = \cup A_i \times I$. Then $E = (F - \cup D_i \times I) \cup \cup D_i \times \{0,1\}$ is a 2-cell whose boundary lies in ∂M – Int F. By incompressibility ∂E bounds a 2-cell E^1 in ∂M – Int F. By irreducibility of M, $E \cup E^1$ bounds a 3-cell in M. With this information it is easy to construct a homeomorphism $f: M \to F \times I$ such that $f(x) = (x, 0)$ for all $x \in F$.

Case 2: $\partial F = \emptyset$. Since $F \neq S^2, P^2$, there is a 2-sided, nonseparating simple closed curve $J \subset F$. By 6.5 we may assume that each component of $\rho^{-1}(J)$ is 2-sided, properly embedded, and incompressible in M. Let A be the component of $\rho^{-1}(J)$ containing J. Since ρ_* is monic, $\pi_1(A) \cong Z$. Thus A is an annulus or a möbius band. The second possibility is ruled out by the fact that a möbius band does not retract to its boundary. Choose a product neighborhood $A \times I$ of A with $\partial A \times I \subset \partial M$. Let $M_1 = M - A \times (0,1)$ and $F_1 = F \cap M_1$.

We claim that $i_*: \pi_1(F_1) \to \pi_1(M_1)$ is an isomorphism. Clearly i_* is monic. To show that i_* is epic, let α be a loop in M_1 based at a point $x_0 \in F_1$. There is a loop β in F based at x_0 homotopic, rel x_0, to α in M. Choose such a β which crosses J minimally. We must have $\beta(I) \subset F - J$. If not, choose a homotopy $h: I \times I \to M$ in general position to A and with $h|I \times 0 = \beta$, $h|I \times 1 = \alpha$, and $h(0 \times I) = h(1 \times I) = x_0$. Then some arc $L \subset h^{-1}(A)$ has both ends in $I \times 0$. Let K be the subinterval of $I \times 0$ with $\partial K = \partial L$. Then $\beta|K$ is homotopic to $h|L$ which is in turn homotopic to a path γ in J. Replace $\beta|K$ by γ to obtain a loop β_1 in F homotopic to β. Since β must cross J in opposite directions at the points $\beta(\partial K)$, β_1 can be homotoped so as to cross J at least two fewer times than does β. This justifies our assertion that $\beta(I) \subset F - J$. We may as well assume that $\beta(I) \subset F_1$. Since A is incompressible, $\pi_1(M_1) \to \pi_1(M)$ is monic. Thus β is homotopic to α in M_1 and the claim that $i_* : \pi_1(F_1) \to \pi_1(M)$ is an isomorphism is established.

By case 1, there is a homeomorphism $f_1 : M_1 \to F_1 \times I$ with $f_1(x) = (x,0)$ for $x \in F_1$. We may assume that $f_1(A \times \{0,1\}) = (J \times \{0,1\}) \times I$. It is now easy to extend f_1 to a homeomorphism $f: M \to F \times I$.

Twisted Bundles

We characterize twisted I-bundles over surfaces. We begin by showing that a free involution on a product $F \times I$ is equivalent to a standard one. Here two groups A_1, A_2 of homeomorphisms of a space X are *equivalent* if they are conjugate in the group of all homeomorphisms: i.e., there is a homeomorphism $g: X \to X$ such that $A_2 = gA_1g^{-1}$. For Y, a subset of X, A_1 and A_2 are *equivalent relative to* Y if there is such a g with $g(Y) = Y$.

10.3. THEOREM. *Let F be a compact, connected 2-manifold different from B^2, S^2, P^2 and let $\tau: F \times I \to F \times I$ be a free involution such that $\tau(F \times \{0,1\}) = F \times \{0,1\}$. Then τ is equivalent relative to $F \times \{0,1\}$ to a free involution $\tau_1 : F \times I \to F \times I$ where either*

(i) $\tau_1(x,t) = (\sigma(x), t)$ $(x \in F,\ t \in I)$, *or*

(ii) $\tau_1(x,t) = (\sigma(x), 1-t)$

for some free involution $\sigma: F \to F$. Thus τ induces an I-bundle structure on $(F \times I)/\tau$ which is trivial if and only if $\tau(F \times 0) = F \times 0$.

PROOF. First consider the case $\tau(F \times 0) = F \times 0$. Let $M = (F \times I)/\tau$ and $p: F \times I \to M$ the covering projection. Let $\overline{F}_i = p(F \times i)$ $(i = 0,1)$. Then $p|F \times i : F \times i \to \overline{F}_i$ is a double covering; so $(p|F \times i)_*(\pi_1(F \times i))$ has index two in $\pi_1(\overline{F}_i)$. But $p_*\pi_1(F \times I)$ has index two in $\pi_1(M)$ and so we must have $i_*\pi_1(\overline{F}_i) = \pi_1(M)$. Since $\overline{F}_i$ is incompressible in M, $i_* : \pi_1(\overline{F}_i) \to \pi_1(M)$ is an isomorphism. Now M is P^2-irreducible, since $\pi_2(M) \cong \pi_2(F \times I) = 0$ and any fake 3-cell in M would lift to $F \times I$ -- which contains no fake 3-cell. By 10.2, there is a homeomorphism $\overline{g}: \overline{F} \times I \to M$ such that $\overline{g}(\overline{F} \times 0) = \overline{F}_0$. The components of $\partial M - \operatorname{Int}(\overline{F}_0 \cup \overline{F}_1)$ are annuli; so we may assume $g(\overline{F} \times 1) = \overline{F}_1$. We have a covering projection $p_1 : F \times I \to \overline{F} \times I$ given by

$$p_1(x, t) = (\pi \overline{g}^{-1} p(x, 0), t)$$

where $\pi : \overline{F} \times I \to \overline{F}$ is projection. The corresponding covering translation $\tau_1 : F \times I \to F \times I$ has the form (i) of the conclusion. Since $\overline{g}_* p_{1*} \pi_1(F \times I) = p_* \pi_1(F \times I)$, $\overline{g} p_1$ is covered by a map $g : F \times I \to F \times I$. One verifies that g is an equivalence relative to $F \times \{0,1\}$ between τ_1 and τ.

Now suppose that $\tau(F \times 0) = F \times 1$. Assume that we can find a surface $G \subset F \times (0, 1)$ such that $\tau(G) = G$ and such that there is an embedding $g_1 : F \times [0, \frac{1}{2}] \to F \times I$ with $g_1(x, 0) = (x, 0)$ and $g_1(F \times \frac{1}{2}) = G$. Note that $\tau(g_1(F \times [0, \frac{1}{2}])) = F \times I - g_1(F \times [0, \frac{1}{2}))$. Define $\sigma : F \to F$ by $\sigma(x) = \pi g_1^{-1} \tau g_1(x, \frac{1}{2})$. Let $g : F \times I \to F \times I$ be given by

$$g(x, t) = \begin{cases} g_1(x, t) & ; \ t \leq \frac{1}{2} \\ \tau g_1(\sigma(x), 1-t); & t \geq \frac{1}{2} \end{cases} .$$

One verifies that $g^{-1} \tau g(x, t) = (\sigma(x), 1-t)$ giving conclusion (ii) of the theorem.

Thus it remains to find the surface G. For this let X be either

(a) a properly embedded 2-cell in $F \times I$ with ∂X not contractible in $\partial(F \times I)$ (hence $\partial F \neq 0$) and with each of $X \cap (F \times 0)$ and $X \cap (F \times 1)$ an arc, or

(b) a properly embedded, 2-sided, incompressible annulus with one boundary component in $F \times 0$ and the other in $F \times 1$.

In either case the arguments of Theorem 10.2 applied to M cut open along X show that there is a homeomorphism $f : F \times I \to F \times I$ with $f(F \times i) = F \times i$ $(i = 0,1)$ and $f(X) = A \times I$ (A an arc or simple closed curve in F). Thus we abbreviate conditions (a) or (b) by saying that X is *vertical* in $F \times I$.

Suppose that:

(*) We are always able to find a vertical 2-cell (if $\partial F \neq \emptyset$) or annulus (if $\partial F = \emptyset$), X, with either $\tau(X) \cap X = \emptyset$ or $X = \tau(X)$. The latter could hold only if X is an annulus.

Then we complete the proof as follows.

Case 1: $\partial F \neq \emptyset$. Choose a collection $X_1, \cdots, X_{2n}$ of pairwise disjoint vertical 2-cells in $F \times I$ such that $\tau(X_i) = X_{n+i}$ (subscripts taken mod 2n). Of all such collections choose one such that the boundary of $F \times I$ cut open along $\cup X_i$ is as simple as possible. Take regular neighborhoods N_i of X_i with $\tau(N_i) = N_{n+i}$. By repeated applications of 10.2, we may assume (after redefining the product structure on $F \times I$ leaving the ends unchanged) that each N_i is "vertical." In particular $N_i \cap C\ell(F \times I - N_i)$ is the union of two vertical 2-cells X_{i0} and X_{i1}.

Let $V_1, \cdots, V_k$ denote the components of $C\ell(F \times I - \cup N_i)$. So $V_i = F_i \times I$ for some surface $F_i \subset F$. We claim that each V_i is a 3-cell. For suppose V_1 is not a 3-cell. If $\tau(V_1) = V_1$, then by (*) applied to V_1, there is a vertical 2-cell X in V_1 with $X \cap \tau(X) = \emptyset$. We may assume $(X \cup \tau(X)) \cap \cup X_{ij} = \emptyset$; if not, choose a component T of $(X \cup \tau(X)) \cap X_{ij}$ which cobounds a 2-cell D in X_{ij} with an arc T' in ∂X_{ij} with $\operatorname{Int} D \cap (X \cup \tau(X)) = \emptyset$. Say $T \subset X$. We modify X in a neighborhood of D to eliminate T. The corresponding change to $\tau(X)$ occurs near $\tau(D)$; so we can preserve the condition $X \cap \tau(X) = \emptyset$. By adding $\{X, \tau(X)\}$ to $\{X_1, \cdots, X_{2n}\}$, we get a contradiction.

If $\tau(V_1) \neq V_1$, then let X be any vertical 2-cell in V_1 (missing $\cup X_{ij}$). Then automatically $X \cap \tau(X) = \emptyset$; so again by adding $\{X, \tau(X)\}$ to $\{X_1, \cdots, X_{2n}\}$, we get a contradiction. This justifies our claim that each V_i is a 3-cell.

Now τ permutes the V_i's, and since τ is fixed point free $\tau(V_i) \neq V_i$. Thus there is an even number, 2q, of the V_i's and we may assume $\tau(V_i) = V_{i+q}$. For $1 \leq i \leq q$. Let $G_i = (F \times \frac{1}{2}) \cap V_i$. For $q+1 \leq i \leq 2q$, let $G_i = \tau(G_{i-q})$. For $1 \leq i \leq n$, N_i meets two V's, say V_j and V_k.

The arcs $G_j \cap N_i$ and $G_k \cap N_i$ form part of the boundary of a 2-cell H_i in N_i which separates $N_i \cap (F \times 0)$ from $N_i \cap (F \times 1)$. For $n+1 \leq i \leq 2n$ put $H_i = \tau(H_{i-q})$. Then $G = \cup G_i \cup \cup H_j$ is the desired surface. The embedding $g_1 : F \times [0, \frac{1}{2}] \to F \times I$ with $g_1(F \times \frac{1}{2}) = G$ can be constructed piecewise as was G. This completes the proof, modulo (*), in the case $\partial F \neq \emptyset$.

Case 2: $\partial F = \emptyset$. By (*) we have a vertical annulus X with either $X \cap \tau(X) = \emptyset$ or $\tau(X) = X$. As before, $X \cup \tau(X)$ is vertical. We choose a "vertical" regular neighborhood N of $X \cup \tau(X)$ with $\tau(N) = N$. Using Case 1 and its proof, we can construct the part of G lying in (the components of) $F \times I - N$. We then extend G across N. If $X \cap \tau(X) = \emptyset$, we extend G across one component of N with an annulus and copy this extension, via τ, in the other component. If $X = \tau(X)$, then the extension of G across N can be chosen to be the inverse image of an annulus or a möbius band in N/τ according as N/τ is a solid Klein bottle or a solid torus.

It remains to verify (*). Let X be a vertical 2-cell (if $\partial F \neq \emptyset$) or annulus (if $\partial F = \emptyset$) in $F \times I$ such that X is in general position with respect to $\tau(X)$. We show how to simplify $X \cap \tau(X)$ by *equivariant surgery*.

Let D be the closure of a component of $\tau(X) - X \cap \tau(X)$. In the applications, D will be a 2-cell or an annulus. Let E be a submanifold of X such that $E \cap \tau(X) = \partial E \cap \tau(X) = D \cap X$ and $\partial E \subset \partial X \cup \partial D$. Suppose that $\tau(D \cap X) \cap (D \cap X) = \emptyset$. Choose a small product neighborhood $D \times [-1, 1]$ of D with $D = D \times 0$, $(D \cap X) \times [-1, 1] \subset X$, and $(D \cap X) \times [-1, 0] \subset E$. Let $E_1 = E \cup (D \cap X) \times [0, 1]$, $D_1 = D \times 1$ and $X_1 = (X - E_1) \cup D_1$. So X_1 is a properly embedded surface. Now $X_1 \cap \tau(X_1) = (X - E_1) \cap \tau(X - E_1) \cup (X - E_1) \cap \tau(D_1) \cup D_1 \cap \tau(X - E_1) \cup D_1 \cap \tau(D_1)$. Now $\tau(D) \cap D = \emptyset$; for $\tau(D) \subset X$ and, by assumption, $\tau(D \cap X) \cap (D \cap X) = \emptyset$. Thus we may assume that D_1 is close enough to D that $\tau(D_1) \cap D_1 = \emptyset$. Also $D_1 \cap \tau(X) = \emptyset$; so $D_1 \cap \tau(X - E_1) = \emptyset$. Thus $(X - E_1) \cap \tau(D_1) =$

$\tau(\tau(X-E_1)\cap D_1) = \emptyset$. Thus $X_1 \cap \tau(X_1) = (X-E_1) \cap \tau(X-E_1) \subset X \cap \tau(X) - (D \cap X)$. So $X_1 \cap \tau(X_1)$ has fewer components than does $X \cap \tau(X)$. It remains to show how to apply this procedure to preserve (a) or (b). Consider the following possibilities:

(1) D is a 2-cell and $D \cap X$ is a simple closed curve J in $\text{Int}\, X \cap \text{Int}\, \tau(X)$. If $\tau(J) = J$, then $D \cup \tau(D)$ is an invariant 2-sphere which must bound a 3-cell $B \subset F \times I$. Now $\tau(B) \neq B$ since τ is fixed point free. Thus we have the contradiction that $S^3 = B \cup \tau(B) = F \times I$. Since τ permutes the components of $X \cap \tau(X)$, $\tau(J) \cap J = \emptyset$; so we can apply the above procedure with E the unique 2-cell in X with $\partial E = J$ to obtain X_1. Since $\partial X_1 = \partial X$, condition (a) (or (b)) is preserved.

(2) D is a 2-cell, $D \cap X$ is an arc J, and D does not meet both $F \times 0$ and $F \times 1$; say $D \cap F \times 1 = \emptyset$. Since τ is fixed point free, $\tau(J) \cap J = \emptyset$. Let E be the closure of the component of $X - J$ which does not meet $F \times 1$. Then E is a 2-cell and $\partial E \subset J \cup \partial X$. We use E to obtain X_1 from X by surgery. Now $\partial(D \cup E)$ is contractible in $F \times I$ and does not meet $F \times 1$; so $\partial(D \cup E)$ is contractible in $\partial(F \times I)$. Thus ∂X is homotopic to ∂X_1 in $\partial(F \times I)$. Now $X_1 \cap F \times 1 = X \cap F \times 1$; so X_1 will satisfy (a) (or (b)) provided $X_1 \cap F \times 0$ is nonempty and connected. The only possible exceptions to the last statement might occur when X is a 2-cell. However, if $X_1 \cap F \times 0 = \emptyset$, we contradict the incompressibility of $\partial(F \times I) - F \times 0$. If $X_1 \cap F \times 0$ is not connected, then $E \cap F \times \{0, 1\} = \emptyset$ and $D \supset \tau(X) \cap F \times 0$. But then $\partial((\tau(X) - D) \cup E)$ is contractible in $F \times I$ and does not meet $F \times 0$; so $\partial(\tau(X))$ is homotopic in $\partial(F \times I)$ to a loop missing $F \times 1$. This contradicts the incompressibility of $\partial(F \times I) - F \times 1$.

(3) D is a 2-cell and $D \cap X$ is an arc J with one end in $F \times 0$ and the other in $F \times 1$ (so X is a 2-cell). Then $\tau(J) \cap J = \emptyset$. There are two choices E', E'' for E giving rise to vertical 2-cells X'_1, X''_1. If both $\partial X'_1$ and $\partial X''_1$ are contractible in $\partial(F \times I)$, then so is ∂X. Thus one of X'_1, X''_1 must satisfy (a).

Repeated applications of the above steps completes the proof of (*) if X is a 2-cell and reduces to one of the following if X is an annulus.

(4) Each component of $X \cap \tau(X)$ is an essential simple closed curve in X. Then let N be an invariant regular neighborhood of $X \cup \tau(X)$. Then two components, X_1, X_2 of $N \cap \overline{F \times I - N}$ are vertical annuli. Either $\tau(X_1) = X_1$ or $\tau(X_1) = X_2$.

(5) Each component of $X \cap \tau(X)$ is a vertical arc. It then follows that $X \cup \tau(X)$ is vertical in $F \times I$. In particular, for a suitable regular neighborhood N of $X \cup \tau(X)$, N is homeomorphic to $(N \cap (F \times 0)) \times I$. The components $X_1, \cdots, X_k$ of $N \cap C\ell(F \times I - N)$ are 2-sided annuli with one boundary component in $F \times 0$ and the other in $F \times 1$. If some X_i is incompressible, we are done; for we may assume $\tau(N) = N$, thus either $\tau(X_i) = X_i$ or $\tau(X_i) \cap X_i = \emptyset$. If each X_i is compressible, the components of $F \times I - N$ are all 3-cells and an I-bundle structure on N/τ extends to an I-bundle structure on $(F \times I)/\tau$. In this final case (*) is not needed, but it still follows by taking the restriction of this bundle to a 2-sided non-separating simple closed curve and pulling back by the covering projection $p : F \times I \to (F \times I)/\tau$.

It is not known, in general, whether a covering space of a 3-manifold which contains no fake 3-cell could contain a fake 3-cell. We will have more to say about this later (13.4, 13.5). At present we need

10.4. LEMMA. *Let* M *be a* P^2*-irreducible 3-manifold and* $p : \tilde{M} \to M$ *a 2-sheeted cover. Then* $\tilde{M}$ *is* P^2*-irreducible.*

PROOF. $\pi_2(\tilde{M}) \cong \pi_2(M) = 0$; so if $\tilde{M}$ is not P^2-irreducible, then $\tilde{M}$ must contain a fake 3-cell. Let C be a homotopy 3-cell in $\tilde{M}$. We may assume that $S = \partial C$ is in general position with respect to $\tau(S)$ where $\tau : \tilde{M} \to \tilde{M}$ is the nontrivial covering translation. If $S \cap \tau(S) = \emptyset$, then $C \cap \tau(C) = \emptyset$; otherwise $C \subset \tau(C)$, or vice versa, and τ would have a fixed point. Thus $p|C$ is an embedding and C is a 3-cell.

If $S \cap \tau(S) \neq \emptyset$, let D be a 2-cell in $\tau(S)$ such that $D \cap S = \partial D$. If $\tau(\partial D) = \partial D$, then $p(D)$ is a projective plane in M. Now, by assumption, $p(D)$ is 1-sided in M. Thus a regular neighborhood N of $p(D)$ is the twisted I-bundle over P^2. By irreducibility ∂N bounds a 3-cell B in M. Thus $M = N \cup B = P^3$ and $\tilde{M} = S^3$. So we may assume that $\tau(\partial D) \cap \partial D = \emptyset$. Thus we can perform equivariant surgery (as in the proof of 10.3). There are two ways of doing this (corresponding to the two 2-cells in S bounded by ∂D) giving rise to 2-spheres S', S'', each of which meets its image under τ in fewer components than does S. These 2-spheres bound homotopy 3-cells C', C'' with one of C, C', C'' contained in the union of the other two. If C is a fake 3-cell, then so is one of C', C'' and we get a contradiction by repeating the process.

10.5. THEOREM. *Let* M *be a compact* 3-*manifold which contains no* 2-*sided* P^2 *and let* $F(\neq B^2, S^2, P^2)$ *be a compact, connected, incompressible surface in* ∂M. *If the index,* $[\pi_1(M) : i_*\pi_1(F)]$, *of* $i_*\pi_1(F)$ *in* $\pi_1(M)$ *is finite, then either*

(i) $\pi_1(M) \cong Z$ *and* $\mathcal{P}(M)$ *is a solid torus or a solid Klein bottle,*

(ii) $[\pi_1(M) : i_*\pi_1(F)] = 1$ *and* $\mathcal{P}(M) = F \times I$ *with* $F = F \times 0$, *or*

(iii) $[\pi_1(M) : i_*\pi_1(F)] = 2$ *and* $\mathcal{P}(M)$ *is a twisted* I-*bundle over a compact* 2-*manifold* $\bar{F}$ *with* F *the associated* 0-*sphere bundle.*

PROOF. For convenience we assume $M = \mathcal{P}(M)$. Then M is P^2-irreducible; otherwise M contains an essential 2-sphere and we cannot have $i_*\pi_1(F)$ of finite index in $\pi_1(M)$.

If $\pi_1(M) = Z$, then by 5.3, M is a solid torus or a solid Klein bottle.

Case (ii) is just a restatement of 10.2.

So we assume that $\pi_1(M) \not\cong Z$ and that $i_*\pi_1(F)$ is a proper subgroup of $\pi_1(M)$. Let $p : \tilde{M} \to M$ be the finite sheeted covering with $p_*\pi_1(\tilde{M}) = i_*\pi_1(F)$. For some component $\tilde{F}_0$ of $p^{-1}(F)$, $p|\tilde{F}_0 : \tilde{F}_0 \to F$ is a homeomorphism. By 10.2, $\mathcal{P}(\tilde{M}) = \tilde{F}_0 \times I$ with $\tilde{F}_0 = \tilde{F}_0 \times 0$. Note that $\partial\tilde{M}$ contains no 2-sphere; so $\mathcal{P}(\tilde{M}) \# \Sigma = \tilde{M}$ for some homotopy 3-sphere Σ.

Let $\tilde{F}_1,\cdots,\tilde{F}_n$ be the other components of $p^{-1}(F)$. If $p|\tilde{F}_i:\tilde{F}_i \to F$ is k_i sheeted, then $\chi(\tilde{F}_i) = k_i\chi(F)$ and $1+k_1+\cdots+k_n = [\pi_1(M):i_*\pi_1(F)]$.

Now $\chi(F) = \chi(\tilde{F}_0) = \chi(\tilde{M}) = \frac{1}{2}\chi(\partial\tilde{M})$, and $\chi(\partial\tilde{M}) = \chi(\tilde{F}_0)+\cdots+\chi(\tilde{F}_n) + \chi(\partial\tilde{M} - \cup\tilde{F}_i) \le \chi(\tilde{F}_0)+\cdots+\chi(\tilde{F}_n) = (1+k_1+\cdots+k_n)\chi(F)$. Thus $\chi(F) \le \frac{1}{2}(1+k_1+\cdots+k_n)\chi(F) \le 0$. If $\chi(F) \ne 0$, then $n = 1$ and $k_1 = 1$. If $\chi(F) = 0$ and F is closed (hence each $\tilde{F}_i$ is closed), then again $n = 1$ (since $\partial\tilde{M}$ has 2-components) and $k_1 = 1$ (since $p_*\pi_1(\tilde{F}_1)$ is conjugate to $p_*\pi_1(\tilde{F}_0)$). If $\chi(F) = 0$ and $\partial F \ne \emptyset$, then ∂M is a compressible torus or Klein bottle and we are back to case (i). So we have $[\pi_1(M):i_*\pi_1(F)] = 2$. By 10.4, $\tilde{M}$ is P^2-irreducible, so $\tilde{M} = \mathcal{P}(\tilde{M}) = \tilde{F}_0 \times I$. It is easy to adjust the product structure so that $\tilde{F}_1 = \tilde{F}_0 \times 1$. Then conclusion (iii) follows from 10.3.

Surface Subgroups of Finite Index

10.6. THEOREM. *Let* M *be a compact* 3-*manifold which contains no* 2-*sided* P^2 *and suppose that* $\pi_1(M)$ *contains a subgroup* G *of finite index which is isomorphic to the fundamental group of some closed surface* ($\ne S^2, P^2$). *Then* $\mathcal{P}(M)$ *is an* I-*bundle over some closed surface. In particular,* $\pi_1(M)$ *is isomorphic to the fundamental group of a closed surface.*

PROOF. We suppose that $M = \mathcal{P}(M)$. As in 10.5, M is P^2-irreducible. Let $p:\tilde{M} \to M$ be the covering of M with $p_*\pi_1(\tilde{M}) = G$. Now $\tilde{M}$ is aspherical. Since $H_2(\pi_1(\tilde{M}))$ is torsion free and $H_3(\pi_1(\tilde{M})) = 0$, $\partial\tilde{M} \ne \emptyset$. Let $\tilde{F}$ be a component of $\partial\tilde{M}$. Then $\tilde{F} \ne S^2, P^2$ and $\tilde{F}$ is incompressible in $\tilde{M}$. Since every subgroup of infinite index in the fundamental group of a 2-manifold is free, $i_*\pi_1(\tilde{F})$ has finite index in $\pi_1(\tilde{M})$. Put $F = p(\tilde{F})$. Then F is a closed incompressible surface in ∂M ($F \ne S^2, P^2$) and $[\pi_1(M):i_*\pi_1(F)] < \infty$. The conclusion now follows from 10.5.

We note that the last part of 10.6 follows from a theorem of Zieschang [115], [116] which asserts that a finitely generated torsion free group which contains a subgroup of finite index isomorphic to the fundamental group of a closed surface is itself isomorphic to the fundamental group of a closed surface.

In this context we note that the deep result of Stallings and Swan [99], [101] that a torsion-free group with a free subgroup of finite index is free also, follows readily for fundamental groups of 3-manifolds with

10.7. THEOREM. *Let* M *be a compact* 3*-manifold with* $\pi_1(M)$ *torsion free. If* $\pi_1(M)$ *contains a nontrivial free subgroup of finite index, then* $\pi_1(M)$ *is free (and the structure of* M *is given by* 5.3).

PROOF. Let G be a free subgroup of finite index in $\pi_1(M)$. Let $\mathcal{P}(M) = M_1 \# \cdots \# M_k$ be the prime factorization of $\mathcal{P}(M)$. Now $\pi_1(M)$ is free if and only if $\pi_1(M_i)$ is free for each i, and $G \cap \pi_1(M_i)$ has finite index in the infinite group $\pi_1(M_i)$. Thus it suffices to prove the theorem in the case M is prime.

If M is closed, then $\pi_2(M) \neq 0$; otherwise the cover of M corresponding to G is a closed aspherical 3-manifold with free fundamental group. This is impossible since free groups have trivial homology in all dimensions greater than one. By the sphere theorem, M contains an essential 2-sphere. Thus (3.13) M is a 2-sphere bundle over S^1 and the conclusion follows in this case.

If $\partial M \neq \emptyset$, then M is P^2-irreducible. If we are able to reduce M to a union of 3-cells by cutting along properly embedded 2-cells, then $\pi_1(M)$ is free. If not, we find a closed, 2-sided incompressible surface ($\neq S^2, P^2$) in M. Thus the (finite sheeted) covering of M corresponding to G contains a closed, 2-sided, incompressible surface. This contradicts the fact that subgroups of free groups are free and completes the proof.

We conclude by noting that if we drop the assumption that M contains no 2-sided P^2 in 10.6 or 10.7, then the hypothesis (and conclusion) still apply to the orientable double cover $\tilde{M}$ of M. However, there may be 2-spheres in $\partial\tilde{M}$ which cover P^2's in ∂M. See 6.10 for a description of such examples.

CHAPTER 11

GROUP EXTENSIONS AND FIBRATIONS

Throughout this chapter we assume that M is a compact 3-manifold satisfying:

(*) There is an exact sequence

$$1 \to N \to \pi_1(M) \overset{\eta}{\to} Q \to 1$$

where N is a finitely generated normal subgroup of $\pi_1(M)$ with infinite quotient group Q.

We prove:

11.1. THEOREM [42]. *Let* M *be a compact 3-manifold satisfying* (*). *Then*:

(1) N *is isomorphic to the fundamental group of a compact 2-manifold.*

(2) *If* $N \not\cong Z$, *then* Q *contains a finite normal subgroup* K *with* Q/K *isomorphic to either* Z *or* $Z_2 * Z_2$.

(3) *If* $N \not\cong Z$ *and* M *contains no 2-sided* P^2, *then either*

(i) *The Poincaré associate* $\mathcal{P}(M)$ *is a fiber bundle over* S^1 *with fiber a compact 2-manifold* F, *or*

(ii) $\mathcal{P}(M)$ *is the union of two twisted* I*-bundles whose intersection,* F, *is the associated* 0*-sphere bundle of each.*

In either case N *is a subgroup of finite index in* $\pi_1(F)$.

(4) *If* $N \not\cong Z$ *and* M *contains a 2-sided* P^2, *then either*

(i) $N \cong Z_2$ *and* $\mathcal{P}(M)$ *is homotopy equivalent to* $P^2 \times S^1$ *or*

(ii) *The orientable double cover of* M *satisfies the conclusion of* (3).

The case $N \cong Z$ leads to a study of Seifert fibered spaces treated in the next chapter; however, Theorems 11.6 and 11.10 give some information in this case.

In (4)(i) we are restricted to the conclusion of homotopy equivalence because of the failure of 10.2 to treat the case $F = P^2$.

Examples for (4)(ii) can be obtained by choosing M_1 to satisfy the conclusion of (3), choosing an involution $\tau : M_1 \to M_1$ with a finite fixed point set (one may not always exist), taking invariant 3-cell neighborhoods $B_1, \cdots, B_n$ of the fixed points and putting $M = (M_1 - \cup \text{Int } B_i)/\tau$.

Algebraic Preliminaries

The remainder of this chapter is devoted to the proof of 11.1. We begin with some group theoretic tools.

11.2. LEMMA. *If* N *is a nontrivial, finitely generated, normal subgroup of a free product* $G_1 * G_2 (G_i \neq 1)$, *then* N *has finite index in* $G_1 * G_2$.

PROOF. Choose 2-complexes $X'_i (i = 1, 2)$ with $\pi_1(X'_i) = G_i$. Join the base point of X'_1 to the base point of X'_2 by an arc L. Let $X = X'_1 \cup L \cup X'_2$ and choose $x_0 \in \text{Int } L$. Then $\pi_1(X) = G_1 * G_2$ and $X - x_0$ has two components. Let X_i denote the closure of the component of $X - x_0$ containing X'_i.

Let $p : \tilde{X} \to X$ be the regular covering space such that $p_* \pi_1(\tilde{X}) = N$. If $p^{-1}(x_0)$ is finite, we are done. So assume $p^{-1}(x_0)$ is infinite. Since N is finitely generated, there is a compact, connected subset C of $\tilde{X}$ such that $i_* : \pi_1(C) \to \pi_1(\tilde{X})$ is epic. Choose $\tilde{x}_0 \in p^{-1}(x_0) - C$. It follows that $\tilde{x}_0$ separates $\tilde{X}$ and that if A is the closure of the component of $\tilde{X} - \tilde{x}_0$ which does not meet C, then $\pi_1(A) = 1$. If $A \cap p^{-1}(x_0) = \tilde{x}_0$, then p must map A homeomorphically to one of X_1, X_2. This contradicts the assumption that $G_i \neq 1$. If A contains more than one point of $p^{-1}(x_0)$, then A contains a component of $p^{-1}(X_1)$ and a component of $p^{-1}(X_2)$. By regularity, each component of $p^{-1}(X_i)$ is simply connected $(i=1,2)$ and each point of $p^{-1}(x_0)$ separates $\tilde{X}$. Thus $N \cong \pi_1(\tilde{X}) = 1$.

11.3. LEMMA. *If* G *is a group whose center,* z(G), *is of finite index, then the commutator subgroup,* [G, G], *of* G *is finite.*

PROOF. Let $\{x_1, \cdots, x_n\}$ be a set of right coset representatives of z(G) in G. For $g \in G$, let $\bar{g}$ denote the unique x_i such that $g \in z(G)x_i$. Note that $\overline{gh} = \overline{\bar{g}h}$.

Now [G, G] is finitely generated (by the set $\{[x_i, x_j] : i, j = 1, \cdots, n\}$) and contains the abelian subgroup $[G, G] \cap z(G)$ of finite index; so it suffices to show that each element of [G, G] has finite order. For this define a function $\phi : G \to z(G)$ by $\phi(g) = \prod_{i=1}^{n} x_i g(\overline{x_i g})^{-1}$. Since each factor $x_i g(\overline{x_i g})^{-1}$ is in z(G), the product is independent of the order. From this it follows that ϕ is a homomorphism. It also follows that $\phi(g) = g^n$ for all $g \in G$; for fix g and note that $\overline{x_1 g}, \cdots, \overline{x_n g}$ is some permutation of $x_1, \cdots, x_n$. Write this permutation as a product of cycles of lengths $\ell_1, \ell_2, \cdots, \ell_k$ respectively. Then $\phi(g) = x_{i_1} g^{\ell_1} x_{i_1}^{-1} \cdots x_{i_k} g^{\ell_k} x_{i_k}^{-1}$. But $x_{i_j} g^{\ell_j} x_{i_j}^{-1} \in z(G)$, so $x_{i_j} g^{\ell_j} x_{i_j}^{-1} = g^{\ell_j}$. Thus $\phi(g) = g^{\ell_1 + \cdots + \ell_k} = g^n$.

Since z(G) is abelian, $[G, G] < \ker \phi$; i.e., $g^n = 1$ for $g \in [G, G]$.

11.4. LEMMA. *If a group* Q *contains an infinite cyclic subgroup of finite index, then* Q *contains a finite normal subgroup* K *with* Q/K *isomorphic to either* Z *or* $Z_2 * Z_2$.

PROOF. By intersecting an infinite cyclic subgroup of finite index with its finitely many distinct conjugates, we obtain a normal, infinite cyclic subgroup L of finite index in Q.

Let $Q_1 = \{x \in Q : xyx^{-1} = y \text{ for all } y \in L\}$ be the centralizer of L in Q. Then $[Q : Q_1] \leq 2$, and $L < z(Q_1)$. By 11.3, $[Q_1, Q_1]$ is finite.

Let $\psi : Q_1 \to Q_1/[Q_1, Q_1]$ be the natural projection. Now $Q_1/[Q_1, Q_1] = Z + T$ where $|T| < \infty$. Since $[Q_1, Q_1]$ is finite, $K_1 = \psi^{-1}(T)$ is precisely the set of elements of finite order in Q_1. Thus K_1 is fully invariant in Q_1, hence normal in Q. Furthermore, Q/K_1 is an extension of Z by Z_2, and must be one of Z, $Z_2 * Z_2$, or $Z + Z_2$ ($Z_2 * Z_2 =$

$\langle a,b : a^2 = b^2 = 1\rangle$ is the semidirect product $\langle x,y : y^2 = 1,\ yxy = x^{-1}\rangle$: put $a = yx$, $b = y$). In the first two cases put $K = K_1$. In the last case let K be the inverse of the Z_2 summand under the projection $Q \to Z + Z_2$.

11.5. LEMMA. *Let* $1 \to N \to G \overset{\eta}{\to} Q \to 1$ *be an exact sequence, where* N *is free of finite rank* $r \geq 2$ *and* Q *is a finitely generated torsion group. Then* G *contains the direct product* $N \times \zeta(N)$ *as a subgroup of finite index where* $\zeta(N) = \{g \in G : [g,x] = 1$ *for all* $x \in N\}$ *is the centralizer of* N *in* G.

PROOF. Since N is normal, each $g \in G$ induces, by conjugation, an automorphism of N. Thus we have a commutative diagram

$$\begin{array}{ccccccccc} 1 & \longrightarrow & N & \longrightarrow & G & \xrightarrow{\ \eta\ } & Q & \longrightarrow & 1 \\ & & & & \Big\downarrow{\scriptstyle\phi} & & \Big\downarrow{\scriptstyle\psi} & & \\ 1 & \longrightarrow & J & \longrightarrow & \mathrm{Aut}\,N & \xrightarrow{\ \theta\ } & \mathrm{Aut}\,N/[N,N] & & \end{array}$$

where $J = \ker\theta$.

The group $\psi(Q)$ is a finitely generated torsion subgroup of $\mathrm{Aut}\,N/[N,N] \cong GL(r,Z)$. It is known (cf. [43; Theorem 2.35]) that such a group is finite. Thus $G_1 = \ker\psi\circ\eta$ has finite index in G. Now $\phi(N)$ is the group, I, of inner automorphisms of N. Thus ϕ induces a homomorphism $\bar{\phi} : G_1/N \to J/I$. Now G_1/N is a torsion group, but it is known [5] that (for a free group) J/I is torsion free. Thus $\phi(G_1) < I$. So for $g \in G_1$, there is some $h \in N$ with $gxg^{-1} = hxh^{-1}$ for all $x \in N$. Thus $g = h(h^{-1}g) \in N\cdot\zeta(N)$ and we have $G_1 = N\cdot\zeta(N)$. Since $r \geq 2$ $N \cap \zeta(N) = 1$; thus $N\cdot\zeta(N) = N \times \zeta(N)$ and the proof is complete.

Bundles

We now turn to some special cases of 11.1. The next combines results of [41] and [96].

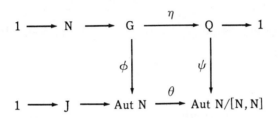

11.6. THEOREM. *Let* M *be a compact 3-manifold satisfying* $(*)$. *In addition, suppose that* Q *is a free group.*

(i) *If* $Q \cong Z$, *then either* $\mathcal{P}(M)$ *is a fiber bundle over* S^1 *with fiber a connected 2-manifold* F *with* $N = i_*\pi_1(F)$ *or* $\mathcal{P}(M)$ *is homotopy equivalent to* $P^2 \times S^1$.

(ii) *If rank* $Q > 1$, *then* $\mathcal{P}(M)$ *is a fiber bundle over a bounded 2-manifold* V *with fiber a circle,* C, *with* $i_*\pi_1(C) = N$.

PROOF. We note that a conclusion analogous to (ii) also holds if Q is the fundamental group of a closed surface (11.10). It is more convenient to postpone consideration of this case.

We assume $M = \mathcal{P}(M)$. Clearly $\pi_1(M) \not\cong Z$, and by 11.2, $\pi_1(M)$ is not a free product. Thus M is irreducible.

Let X be a wedge of r circles $X_1, \cdots, X_r$ ($r = \text{rank } Q$) with common base point x_0. Choose $x_i \in X_i - x_0$. By identifying $\pi_1(X)$ with Q and applying 6.5, we get a map $f: M \to X$ such that $f_* = \eta : \pi_1(M) \to Q$ and such that each component of $f^{-1}(\{x_1, \cdots, x_r\})$ is a properly embedded, 2-sided imcompressible surface in M. We denote these components by $F_1, \cdots, F_n$ and denote the components of M cut open along $\cup F_i$ by $R_1, \cdots, R_m$.

We construct a connected 1-complex Γ in M "dual to" $\cup F_i$; i.e., Γ has one vertex in each R_j and one edge for each F_i, $\Gamma \cap R_j$ is a tree and $\Gamma \cap F_i$ is a single point at which an edge of Γ crosses F_i transversely. We define a retraction $\rho : M \to \Gamma$ by defining ρ to be projection to the I-factor on an appropriate product neighborhood $F_i \times I$ of each F_i and extending over each R_j using the contractibility of $R_j \cap F$. In particular, $\rho^{-1}(\Gamma \cap F_i) = F_i$.

For any loop $\alpha : (I, \partial I) \to (M, y_0)$ $(y_0 \in f^{-1}(x_0) \cap \Gamma)$, the element $f_*([\alpha]) \in \pi_1(X, x_0)$ can be written as a word in the standard free generators for $\pi_1(X, x_0)$ by observing the oriented intersections of α with the F_i's. This intersection pattern is the same for $\rho \circ \alpha$. Thus the following diagram is commutative.

$$\begin{array}{ccc} \pi_1(M) & \xrightarrow{\ f_*\ } & \pi_1(X) \\ & {\scriptstyle \rho_*}\searrow \qquad \nearrow{\scriptstyle (f|\Gamma)_*} & \\ & \pi_1(\Gamma) & \end{array}$$

Since ρ_* is epic $\rho_*(\ker f_*) = \ker(f|\Gamma)_*$; thus $\ker(f|\Gamma)_*$ is finitely generated. But $\ker(f|\Gamma)_*$ has infinite index in the free group $\pi_1(\Gamma)$. By 11.2 (in the case rank $\pi_1(\Gamma) > 1$, and common sense otherwise) $\ker(f|\Gamma)_* = 1$. Thus $(f|\Gamma)_*$ is an isomorphism and $\ker \rho_* = N$.

Let T be a maximal tree in Γ; so $\Gamma - T$ has exactly r edges. Let $F_1, \cdots, F_r$ be the components of $f^{-1}(\{x_1, \cdots, x_r\})$ dual to these edges. Let $R = M - \bigcup_{j=1}^{r} F_j \times (0,1)$ be M cut open along $F_1 \cup \cdots \cup F_r$. Put $F_{ij} = F_i \times j \subset \partial R (j=0,1)$. Note that R is connected and that $\rho(R) \subset T$. Thus $i_*\pi_1(R) < N$. Observe there is a map $g: R \to M$ such that $g|R - \cup F_{ij}$ is a homeomorphism with image $M - \cup F_i$ and such that g takes each of F_{i0} and F_{i1} homeomorphically to F_i.

Let $p: \tilde{M} \to M$ be the regular covering space with $p_*\pi_1(\tilde{M}) = N$. Since g is homotopic to the inclusion of R into M, g lifts to a map $\tilde{g}: R \to \tilde{M}$. It is easy to check that $\tilde{g}$ is an embedding, that $\tilde{M}$ is the union of the translates of $\tilde{g}(R)$, and that $p^{-1}(F_i)$ is the union of the translates of $\tilde{g}(F_{i0})$.

Now $\pi_1(\tilde{M})$ is finitely generated; so there is a compact, connected subset K of $\tilde{M}$ with $i_*: \pi_1(K) \to \pi_1(\tilde{M})$ epic. For each i each component of $p^{-1}(F_i)$ separates $\tilde{M}$. This is clear for a component $\tilde{F}_i$ disjoint from K (one exists since $\tilde{M}$ is noncompact) since a loop meeting $\tilde{F}_i$ transversely in a single point could not be homotopic to a loop in K. By regularity it holds for all components.

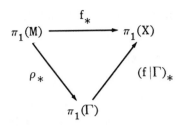

Choose a covering translation τ such that $\tau\tilde{g}(R) \cap K = \emptyset$. By the above for some (i,j), say $(i,j) = (1,0)$, $\tau\tilde{g}(F_{10})$ separates K from the rest of $\tau\tilde{g}(R)$. It follows that $i_*: \pi_1(\tau\tilde{g}(F_{10})) \to \pi_1(\tau\tilde{g}(R))$ is epic and hence is an isomorphism. For let α be a loop in $\tau\tilde{g}(R)$ based at a point of $\tau\tilde{g}(F_{10})$. Then there is a homotopy $h: I \times I \to \tilde{M}$ such that $h(s,0) = \alpha(s)$, $h(0,t) = h(1,t) \notin \tau\tilde{g}(R)$ for $t > 0$ and $h(s,1) \in K$. Using the incompressibility of each $\tau\tilde{g}(F_{k\ell})$ to eliminate simple closed curves in $h^{-1}(\tau\tilde{g}(\cup F_{k\ell}))$, we find that some component of $h^{-1}(\tau\tilde{g}(F_{10}))$ is an arc which, together with $I \times 0$, bounds a 2-cell D in $I \times I$ with $h(D) \subset \tau\tilde{g}(R)$. This establishes our claim.

We use 10.2 in the case $F_{10} \neq P^2$ to conclude that R is homeomorphic to $F_{10} \times I$ with F_{10} corresponding to $F_{10} \times 0$. If $F_{10} = P^2$, then by 9.6, R is homotopy equivalent to $P^2 \times I$.

Now if $Q = Z$, then $r = 1$ and we may assume that F_{11} corresponds to $F_{10} \times 1$. Thus M is a fiber bundle over S^1 with fiber F_{10} (if $F_{10} \neq P^2$) or is homotopy equivalent to $P^2 \times S^1$ (if $F_{10} = P^2$). Furthermore, $\tilde{M}$ is homeomorphic to $\tilde{g}(F_{10}) \times R^1$ (homotopy equivalent to $P^2 \times R^1$); so $i_*\pi_1(F_{10}) = p_*\pi_1(\tilde{M}) = N$. This gives conclusion (i).

If $r > 1$, then ∂R contains $2r \geq 4$ F_{ij}'s and each is incompressible. Since $R = F_{10} \times I$, the only possibility is that the F_{ij}'s are parallel annuli. Thus we may also represent $R = B^2 \times S^1$ where each F_{ij} is the product of an arc in ∂B^2 with S^1. We may adjust the product structure so that g identifies each fiber (pt. $\times S^1$) in F_{i0} with a fiber in F_{i1}. Thus M is an S^1 bundle over a bounded 2-manifold. By construction $N = i_*\pi_1(C)$ where C is any fiber. This gives conclusion (ii) and completes the proof.

Observe that 11.6 (i) applies to 3-manifolds M with $H_1(M) \cong Z$; e.g., *knot spaces* (the complement, in S^3, of an open regular neighborhood of a p.l. embedded simple closed curve):

11.7. COROLLARY. *Let* M *be a compact* 3-*manifold with* $H_1(M) \cong Z$. *If the commutator subgroup* $[\pi_1(M), \pi_1(M)]$ *is finitely generated, then* $\mathcal{P}(M)$ *is homotopy equivalent to* $P^2 \times S^1$ *or* $\mathcal{P}(M)$ *is a fiber bundle over* S^1 *with fiber a compact* 2-*manifold* F *and* $[\pi_1(M), \pi_1(M)] = i_*\pi_1(F)$.

PROOF. 11.6 applies directly if $[\pi_1(M), \pi_1(M)] \neq 1$. If $[\pi_1(M), \pi_1(M)] = 1$, then $\mathcal{P}(M)$ is a 2-sphere or disk bundle over S^1.

We point out that for M a knot space it is known [15], [85] that if $[\pi_1(M), \pi_1(M)]$ is finitely generated, then $|\Delta(0)| = 1$ ($\Delta(t)$ the Alexander polynomial). The converse is true for the space of an alternating knot [74].

11.8. THEOREM. *Let M be a compact 3-manifold satisfying (*) and which contains no 2-sided P^2. If $Q \cong Z_2 * Z_2$, then $\mathcal{P}(M) = M_1 \cup M_2$ where $M_1 \cap M_2 = F$ is a properly embedded 2-sided incompressible surface, M_i is a twisted I-bundle over some surface with F the corresponding 0-sphere bundle, and $N = i_*\pi_1(F)$.*

PROOF. Assume $M = \mathcal{P}(M)$. By 11.2 and hypothesis M is P^2-irreducible. Let X_1 and X_2 be disjoint copies of P^3. Join a point of X_1 to a point of X_2 by an arc L to obtain a complex $X = X_1 \cup L \cup X_2$. Since $\pi_2(X) = 0$, there is a map $f: M \to X$ such that $f_* = \eta: \pi_1(M) \to Q = \pi_1(X)$. Pick $x_0 \in \operatorname{Int} L$. By 6.5 we may assume that each component of $f^{-1}(x_0)$ is a properly embedded, 2-sided, incompressible surface. Of all choices for f we assume that $f^{-1}(x_0)$ has a minimal number of components which we denote $F_1, \cdots, F_n$. Let $R_1, \cdots, R_m$ be the closures of the components of $M - \cup F_i$. Note that each F_i separates M.

Now Q contains an infinite cyclic subgroup, $\tilde{Q}$, of index two. Let $p: \tilde{M} \to M$ be the two sheeted cover such that $p_*\pi_1(\tilde{M}) = \eta^{-1}(\tilde{Q})$.

We have an exact sequence

$$1 \to \tilde{N} \to \pi_1(\tilde{M}) \overset{\eta}{\to} \tilde{Q} \to 1$$

where $p_*|\tilde{N}: \tilde{N} \to N$ is an isomorphism.

By 10.4, $\tilde{M}$ is P^2-irreducible. In particular, $\tilde{M} = \mathcal{P}(\tilde{M})$. By 11.6, $\tilde{M}$ is a fiber bundle over S^1 with fiber a surface $\tilde{F} (\neq P^2)$ and $i_*\pi_1(\tilde{F}) = \tilde{N}$. There is a covering projection $q: \tilde{F} \times R^1 \to \tilde{M}$ with $q_*\pi_1(\tilde{F} \times R^1) = \tilde{N}$.

For any i, $f(F_i) = x_0$; so any component, $\overline{F}_i$, of $q^{-1}p^{-1}(F_i)$ projects homeomorphically to F_i. Let A and B denote the closures of the components of $\tilde{F} \times R^1 - \overline{F}_i$. At least one of these, say A, is noncompact, and thus contains $\tilde{F} \times t$ for some $t \in R^1$. Since $\overline{F}_i$ is incompressible, it follows (as in 11.6) that $\pi_1(\overline{F}_i) \to \pi_1(B)$ is an isomorphism. If B is compact, then by 10.2, $B = \overline{F}_i \times I$. We must have $\partial B - \overline{F}_i \subset \partial\tilde{F} \times R^1$ (hence $\overline{F}_i$ is an annulus). Then $pq|B$ is an embedding, and F_i is parallel to an annulus in ∂M. But then F_i could be eliminated from $f^{-1}(x_0)$ by a homotopy — contrary to our minimality assumptions. Thus B is also noncompact; so by the above argument $\pi_1(\overline{F}_i) \to \pi_1(A)$ is also an isomorphism. Thus, by projecting down, $i_*\pi_1(F_i) = N$.

Now for each j, $f(R_j)$ is in one of X_1, X_2; so $|f_* i_* \pi_1(R_j)| \leq 2$. Hence for any i with $F_i \subset \partial R_j$, $[\pi_1(R_j) : i_*\pi_1(F_i)] \leq 2$. By 10.5, either

(a) $R_j = F_i \times I$ (for any i with $F_i \subset \partial R_j$), or

(b) R_j is a twisted I-bundle with any $F_i \subset \partial R_j$ the corresponding 0-sphere bundle.

We note that any R_j of type (a) must contain at least two F_i's; otherwise some F_i is parallel to a surface in ∂M and could be eliminated.

If $N \neq Z$, then each R_j of type (a) contains exactly two F_i's. Putting these R_j together end to end, we eventually come to one of type (b) and must stop. Proceeding in the other direction we again come to an R_j of type (b) and stop. Thus there are two R_j's of type (b) joined by a product and we have the desired conclusion.

If $N = Z$, we could in principle, have more R_j's of type (b). However, we can describe $\pi_1(M)$ as an amalgamated tree product of infinite cyclic groups:

$$\pi_1(M) = \langle z_1, \cdots, z_k : z_1^2 = z_2^2 = \cdots = z_k^2 \rangle$$

where k is the number of R_j's of type (b). Since N corresponds to the amalgamating subgroup, we have $\pi_1(M)/N \cong Z_2 * \cdots * Z_2$ (k-factors). Thus $k = 2$ and the conclusion follows as above.

11.9. LEMMA. *If* M *satisfies* (*) *and* $N \not\cong Z$, *then* Q *contains an element of infinite order.*

PROOF. First suppose N is free (of rank ≥ 2). If Q is a torsion group, then by 11.5, $\pi_1(M)$ contains $N \times \zeta(N)$ as a subgroup of finite index. Now $\eta|\zeta(N):\zeta(N) \to Q$ is monic, so $\zeta(N)$ is a torsion group. Since N has infinite index in $\pi_1(M)$, $\zeta(N)$ is infinite. In particular, $\zeta(N)$ contains an element of order greater than two (if every element has order two, $\zeta(N)$ is abelian and hence finite). By 9.8, $\pi_1(M)$ is a nontrivial free product. This contradicts 11.2 and proves the lemma for the case N free.

So suppose N is not free. Let $p:\tilde{M} \to M$ be the regular covering space with $p_*\pi_1(\tilde{M}) = N$ and with Q as group of covering translations.

Since N is not free, some finitely generated subgroup $1 \neq A$ of N is neither infinite cyclic nor a nontrivial free product. Following the proof of 8.2, there is a compact 3-manifold $R \subset \text{Int}\,\tilde{M}$ with incompressible boundary such that $i_*\pi_1(R)$ contains some conjugate of A.

Since N is finitely generated, there is a compact, connected set $K \subset \tilde{M}$ with $i_*:\pi_1(K) \to \pi_1(\tilde{M})$ an epimorphism. Note that each closed surface in $\text{Int}\,\tilde{M}$ separates $\tilde{M}$; this is clear for a surface missing K. Since Q is infinite, any surface can be translated off K; so it holds for all surfaces.

We may assume $R \cap K = \emptyset$. Let S be the component of ∂R separating K from the rest of R. Let U be the closure of the component of $\tilde{M} - S$ which misses K. Since S is incompressible and $\pi_1(K) \to \pi_1(\tilde{M})$ is epic, the argument in 11.6 shows that $\pi_1(S) \to \pi_1(U)$ is an isomorphism. Since $R \subset U$, $\pi_1(S) \neq 1$.

Now Q is infinite, so there is an infinite set $\{\tau_1, \tau_2 \cdots\}$ of covering translations with $\tau_i(S) \cap \tau_j(S) = \emptyset$ for $i \neq j$. Put $S_i = \tau_i(S)$ and $U_i = \tau_i(U)$. If for some $i, j (i \neq j)$ $U_i \subset U_j$, let $\tau = \tau_j\tau_i^{-1}$. Then $\tau(U_i) = U_j \supsetneq U_i$; hence $U_i \subsetneq \tau(U_i) \subsetneq \tau^2(U_i) \subsetneq \cdots$ and τ has infinite order.

Choose a component of $M - \bigcup_{i=1}^{\infty} S_i$ whose closure, W, contains at least two S_i's. If for some i, j with $S_i \cup S_j \subset \partial W$, we have $U_i \cap W = S_i$ and $W \subset U_j$, then $U_i \subset U_j$ and, as above, we obtain an element of infinite order. So we assume either

(i) For all i with $S_i \subset \partial W$, $U_i \cap W = S_i$, or

(ii) For all i with $S_i \subset \partial W$, $W \subset U_i$.

In case (ii), let $V_i = C\ell(\tilde{M} - U_i)$. Then for some i, j, $V_i \subset U_j$. Since $\pi_1(S_j) \to U_j$ is an isomorphism, it follows that $\pi_1(S_i) \to \pi_1(V_i)$ is an isomorphism, for this i -- hence for every i. Interchanging the roles of the U_i's and V_i's reduces to (i).

So we assume (i). If for some k, $S_k \cap W = \emptyset$, then for some i with $S_i \subset \partial W$ either $U_k \subset U_i$ or $U_i \subset U_k$. Thus we may assume that $U_i \cap W = S_i$ for every i. Thus ∂W contains infinitely many closed, nonsimply connected surfaces. However, $\pi_1(W) \to \pi_1(\tilde{M})$ is an isomorphism; so $\pi_1(W)$ is finitely generated. Thus $H_1(W; Z_2)$ is finitely generated as is $H_1(W - X; Z_2)$ where X is a finite 1-complex in Int W carrying a set of generators for $\pi_1(W)$. Thus $\ker(H_1(\cup S_i; Z_2) \to H_1(W - X; Z_2)) \neq 0$. A nontrivial element z of this kernel can be represented $z = z_1 + \cdots + z_k$ where z_i is carried by a nonseparating simple closed curve in S_{n_i} and $n_i \neq n_j$ for $i \neq j$. So z bounds a 2-chain C in $W - X$. Some loop, J, in S_{n_1} has (mod 2) intersection number 1 with the relative 2-cycle determined by C. However J is freely homotopic in W to a loop (in X) missing C. This contradicts invariance of intersection numbers and completes the proof.

We now proceed with the

Proof of Theorem 11.1

If $N = Z$, the only part of the theorem which applies is trivial. So assume that $N \neq Z$. Also assume that $M = \mathcal{P}(M)$.

By 11.9, Q contains an infinite cyclic subgroup Q_1. Let $p_1 : M_1 \to M$ be the covering space with $p_{1*}\pi_1(M_1) = \eta^{-1}(Q_1)$. We have an exact sequence

$$1 \to N_1 \to \pi_1(M_1) \to Q_1 \to 1$$

where $p_{1*}|N_1 : N_1 \to N$ is an isomorphism. Being an extension of one finitely generated group by another, $\pi_1(M_1)$ is finitely generated. By 11.2,

$\pi_1(M_1)$ is not a free product and $\pi_1(M_1) \not\cong Z$. Thus by the proof of 8.2, there is a compact 3-manifold R in Int M_1 with incompressible boundary and with $\pi_1(R) \to \pi_1(M_1)$ an isomorphism. By 11.6, $\mathcal{P}(R)$ is a fiber bundle over S^1 or is homotopy equivalent to $P^2 \times S^1$. In any case, there is a properly embedded, 2-sided, nonseparating, incompressible surface $F_1 \subset R$ with $i_*\pi_1(F_1) = N_1 (\cong N)$. This proves part (1).

Now let $p: \tilde{M} \to M$ be the regular covering with $p_*\pi_1(\tilde{M}) = N$. There is a regular covering projection $p_2: \tilde{M} \to M_1$ with $p = p_1 \circ p_2$. Identify Q with the group of covering translations of $p: \tilde{M} \to M$, and let τ be a generator of Q_1. Now $\tilde{R} = p_2^{-1}(R)$ is connected and $\tau|\tilde{R}$ generates the group of covering translations of $p_2|\tilde{R}: \tilde{R} \to R$. We can write $\tilde{R} = \bigcup_{i=-\infty}^{\infty} X_i$ where X_i is homeomorphic to R cut open along F_1, $\tilde{F}_i = X_{i-1} \cap X_i$ is a component of $p_2^{-1}(F_1)$, and $\tau(X_i) = X_{i+1}$ for all i. The components of $\partial\tilde{R}$ (if any) are open annuli ($S^1 \times R^1$) or 2-spheres. Since $\pi_1(\tilde{R}) \to \pi_1(\tilde{M})$ is an isomorphism, the components of $\tilde{M} - \tilde{R}$ have infinite cyclic or trivial fundamental group. Now $\pi_1(\tilde{F}_0) \cong N \neq 1, Z$. So for every $\sigma \in Q$, $\sigma(\tilde{F}_0) \cap \tilde{R} \neq \emptyset$. Now X_0 is compact, so there are only finitely many elements, say $\sigma_1, \cdots, \sigma_n$ in Q with $\sigma_i(\tilde{F}_0) \cap X_0 \neq \emptyset$. For any $\sigma \in Q$, $\sigma(\tilde{F}_0) \cap X_k \neq \emptyset$ for some k. Thus $\tau^{-k}\sigma(\tilde{F}_0) \cap X_0 \neq \emptyset$ and $\tau^{-k}\sigma = \sigma_i$ for some i. This shows that $\sigma_1, \cdots, \sigma_n$ form a set of coset representatives for Q_1 in Q; hence Q_1 has finite index in Q. By 11.4, Q contains a finite normal subgroup K with Q/K isomorphic to either Z or $Z_2 * Z_2$. This proves part (2).

Now note that $\eta^{-1}(K)/N \cong K$; so $\eta^{-1}(K)$ is finitely generated and not infinite cyclic. We have an exact sequence

$$1 \to \eta^{-1}(K) \to \pi_1(M) \to Q/K \to 1 .$$

If M contains no 2-sided P^2, we apply 11.6 or 11.8 (according as $Q/K = Z$ or $Z_2 * Z_2$) to obtain part (3).

Finally, if M contains a 2-sided P^2, let $\bar{p}: \bar{M} \to M$ be the orientable double cover. We have an exact sequence

$$1 \to \overline{N} \to \pi_1(\overline{M}) \to \overline{Q} \to 1$$

where $\overline{N}$ and $\overline{Q}$ are isomorphic to subgroups of index at most two in N and Q respectively. Thus we can apply part (3) to $\overline{M}$ unless $\overline{N} = 1, Z$. If $\overline{N} = Z$, then N is one of $Z, Z+Z_2$, $Z_2 * Z_2$. The first case is prohibited by assumption, the other two by part (1) since neither is the fundamental group of a surface. If $\overline{N} = 1$, then $N = Z_2$. But any normal Z_2 subgroup is central. Thus N together with any element of infinite order in $\pi_1(M)$ generate a subgroup isomorphic to $Z+Z_2$. By 9.12 and the fact that $\pi_1(M)$ is not a free product, $\pi_1(M) \cong Z+Z_2$. Thus M is homotopy equivalent to $P^2 \times S^1$. This gives part (4) and completes the proof.

The following theorem extends 11.6.

11.10. THEOREM. *Let* M *be a compact* 3-*manifold satisfying* (*). *In addition, suppose* $Q \cong \pi_1(T)$ *where* T *is a closed surface* ($\neq S^2, P^2$). *Then* $\mathcal{P}(M)$ *is an* S^1-*bundle over* T *with* N *the fundamental group of the fiber.*

PROOF. Assume $M = \mathcal{P}(M)$. Since Q does not satisfy 11.1(2), we must have $N \cong Z$. By 11.2, M is irreducible, and since Q is torsion free, M is P^2-irreducible.

We first show that M is closed. For this it suffices to assume that M is orientable (since the orientable double cover also satisfies the hypothesis). Since N is central in a subgroup of index at most two in $\pi_1(M)$, we also assume (by passing to a double cover) that N is central in $\pi_1(M)$. Suppose $\partial M \neq \emptyset$, and let $S_1, \cdots, S_k$ be the components of ∂M. Each S_i is incompressible. If for some i, $N \cap \pi_1(S_i) = 1$, then $N \cdot \pi_1(S_i)$ is the fundamental group of a closed, aspherical 3-manifold $(S_i \times S^1)$. By 9.11, M is closed.

So we suppose that for every i, $N_i = N \cap \pi_i(S_i) \neq 1$. Then S_i is a closed, orientable surface whose fundamental group has nontrivial center.

Thus S_i is a torus. Since $\eta(\pi_1(S_i))$ is torsion free, N_i is generated by a simple closed curve L_i in S_i. For each i we attach a solid torus to M along S_i with L_i bounding a meridional disk. This gives a closed orientable 3-manifold M_1 containing M. $\overline{N} = \ker(i_* : \pi_1(M) \to \pi_1(M_1))$ is the smallest subgroup containing each N_i. If $\overline{N} \neq N$, then $\pi_1(M_1)$ contains the finite group $N/\overline{N}$. By 9.8, $\pi_1(M_1)$ can be written as a free product with $N/\overline{N}$ conjugate to a subgroup of one of the factors. This is impossible since $N/\overline{N}$ is normal in $\pi_1(M_1)$. Hence $N = \overline{N}$; so $\pi_1(M_1) \cong \pi_1(T)$. This contradicts 10.6, since M_1 is closed. Thus we must have $\partial M = \emptyset$.

Now let $f : M \to T$ be a map such that $f_* = \eta : \pi_1(M) \to Q = \pi_1(T)$. Let J be a 2-sided nonseparating simple closed curve in T. By 6.5, we may assume that the components $F_1, \cdots, F_n$ of $f^{-1}(J)$ are closed, 2-sided, incompressible surfaces in M. Let $R_1, \cdots, R_m$ be the components of M cut open along $\cup F_i$. For each j, we have an exact sequence

$$1 \to N_j \to \pi_1(R_j) \to Q_j \to 1$$

where $N_j = N \cap \pi_1(R_j)$. Since $f(R_j) \subset T-J$, $Q_j = f_*(\pi_1(R_j))$ is free. $Q_j \neq 1$; otherwise, $\pi_1(R_j) \cong Z$ and R_j could not have incompressible boundary.

Now for each i, $f(F_i) \subset J$; so $N \cap \pi_1(F_i) \neq 1$. From this it follows that $N_j \neq 1$ for every j.

By 11.6 (part ii), there is an S^1-bundle projection $g_j : R_j \to T_j$, for some bounded surface T_j, with $\ker g_{j*} = N_j$. Now for each i, F_i has fibrations induced by the two (perhaps the same) R_j's which lie on opposite sides of F_i. Let C_1 and C_2 be fibers from these fibrations. Each of C_1, C_2 is a power of a generator of N; thus some power of C_1 is homotopic to a power of C_2 in M -- hence in F_i. Since C_1 and C_2 are simple, they are homotopic, hence isotopic (2.10), in F_i. Thus we can piece the T_j's together to get a closed surface $\overline{T}$ and extend the g_j's to an S^1-bundle projection $g : M \to \overline{T}$. By construction, $\ker g_* < N$.

Since $\pi_1(\overline{T})$ is torsion free, $\ker g_* = N$. Thus $\pi_1(\overline{T}) \cong \pi_1(T)$ and so $\overline{T}$ is homeomorphic to T.

11.11. EXERCISE. *Let* M *be a compact 3-manifold with* $\pi_1(M)$ *infinite. If* $\pi_1(M)$ *is a nontrivial direct product, show that* $\mathcal{P}(M)$ *is a cartesian product* $F \times S^1$ *for some compact surface* F, *or* $\mathcal{P}(M)$ *is homotopy equivalent to* $P^2 \times S^1$.

We conclude by noting that the bundle structure on a surface bundle over S^1 is by no means unique. For examples of this, let F_g be a closed orientable surface of genus g. Then

$$\pi_1(F_g \times S^1) = \langle a_1, b_1, \cdots, a_g, b_g, t : [a_i, t] = [b_i, t] = 1,\ [a_1, b_1] \cdots [a_g, b_g] = 1 \rangle .$$

For any integer $n > 0$, define a homomorphism $\phi_n : \pi_1(F_g \times S^1) \to Z$ by $\phi_n(a_1) = 1$, $\phi_n(a_2) = \cdots = \phi_n(a_g) = \phi_n(b_1) = \cdots = \phi_n(b_g) = 0$, $\phi_n(t) = n$.

11.12. EXERCISE. *Show (say, using the Reidermeister-Schreier rewriting process* [63]) *that* $\ker \phi_n \cong \pi_1(F_{n(g-1)+1})$. *Deduce that* $F_g \times S^1$ *is also a fiber bundle over* S^1 *with fiber* $F_{n(g-1)+1}$.

This can also be established by geometric techniques [50].

CHAPTER 12
SEIFERT FIBERED SPACES

We continue the investigation of the preceding chapter by considering those compact 3-manifolds, M, whose fundamental group contains an infinite cyclic normal subgroup. With the assumption that M contains an incompressible surface, we show in 12.7, 12.8, 12.9 that M is, or is closely related to, a Seifert fibered space. If $\pi_1(M)$ has non-trivial (infinite) center, then $\pi_1(M)$ contains an infinite cyclic normal subgroup; so this includes a description of 3-manifolds whose fundamental group has nontrivial center. If the center is not cyclic (and is finitely generated), we obtain a more explicit description of M (12.10). The techniques also yield an explicit description of the quotients of $S^1 \times S^1 \times S^1$ by free cyclic actions.

Seifert fibered spaces are treated in detail in [91] where a complete set of invariants for them is developed. We will not pursue this development; since, with a few exceptions, they fall in the category of 3-manifolds (treated in Chapter 13) which are completely determined by their fundamental group system. We do need a definition and some elementary properties.

A 3-manifold M is a *Seifert fibered space* if M is the union of a collection $\{C_\alpha\}$ of pairwise disjoint simple closed curves (called fibers) such that for each α, there is a closed neighborhood V of C_α which is a solid torus and a covering map $p : B^2 \times S^1 \to V$ satisfying

(i) p maps each $x \times S^1 (x \in B^2)$ to some C_β (hence V is a union of fibers),

(ii) $p^{-1}(C_\alpha)$ is connected, and

(iii) the group of covering translations is generated by $\tau_{n,m}$ for some pair (n, m) of relatively prime integer where

$$\tau_{n,m}(re^{i\theta}, e^{i\phi}) = \left(re^{i\left(\theta + 2\frac{m}{n}\pi\right)}, e^{i\left(\phi + \frac{2\pi}{n}\right)}\right).$$

If $|n| = 1$, then p is an embedding and we say that C_α is a regular fiber. If $|n| > 1$, then by (ii), $C_\alpha = p(0 \times S^1)$, and for $x \neq 0$, p maps $x \times S^1$ homeomorphically to a fiber which crosses the meridional disk $p(B^2 \times 1)$ n times and wraps m times meridionally about C_α. That is, there are standard generators μ, λ for $\pi_1(\partial V)$ with μ nullhomotopic in V and λ generating $\pi_1(V)$ such that for $\beta \neq \alpha$, C_β represents the element $\mu^m \lambda^n$ of $\pi_1(V - C_\alpha) \cong \pi_1(\partial V)$. In this case we say C_α is a singular fiber of type (n, m). We will always assume that $n > 0$, and since $\pm m$ may be changed modulo n, that $0 \leq m \leq n/2$.

Since every other fiber in some neighborhood of a singular fiber is regular, it follows that if M is compact, there can be but finitely many singular fibers. Furthermore, by invariance of domain, ∂M is a union of regular fibers.

Let S be the orbit space obtained from M by identifying each C_α to a point and let $\eta: M \to S$ be the identification map. For each fiber C_α, the map $\eta \circ p|B^2 \times 1 : B^2 \times 1 \to S$ maps $B^2 \times 1$ onto a closed neighborhood of the point $\eta(C_\alpha)$. This map is an embedding if C_α is regular and is equivalent to the projection onto the orbit space of $B^2 \times 1$ under a periodic rotation otherwise. In either case $\eta(p(B^2 \times 1)$ is a 2-cell. Thus S is a 2-manifold called the associated *Seifert surface*. If M has singular fibers $C_{\alpha_1}, \cdots, C_{\alpha_q}$, let $B_1, \cdots, B_q$ be pairwise disjoint 2-cell neighborhoods of $\eta(C_{\alpha_1}), \cdots, \eta(C_{\alpha_q})$ respectively in Int S. Let $S^* = S - \cup \text{Int } B_i$ and $M^* = \eta^{-1}(S^*)$. Then $\eta|M^* : M^* \to S^*$ is a fiber bundle map. We may suppose that $\partial S^* \neq \emptyset$ (if necessary, pretend that at least one fiber is singular). Thus the bundle has a cross section. We may choose "standard generators" c_i, t_i for the torus $\eta^{-1}(\partial B_i)$ where t_i is represented by a (regular) fiber and c_i by a boundary curve of the cross section. Recall that we also have generators μ_i, λ_i for $\eta^{-1}(\partial B_i)$ where μ_i is represented by the boundary of a meridional disk of the solid torus $\eta^{-1}(B_i)$ and $t_i = \mu_i^{m_i}\lambda_i^{n_i}$ where C_{α_i} is of type (m_i, n_i). Thus $c_i = \mu_i^{r_i}\lambda_i^{s_i}$ where $s_i m_i - r_i n_i = 1$. Thus $\mu_i = t_i^{s_i} c_i^{-n_i}$. So $\pi_1(M)$ can be computed from $\pi_1(M^*)$ by adding the relations $t_i^{s_i} c_i^{-n_i} = 1$. The t_i's can be oriented so as to be freely homotopic in M^*, so by properly referring them to a common base point, they represent a single element. We have

12.1. THEOREM. *Let* M *be a compact Seifert fibered space with* q *singular fibers and whose Seifert surface,* S, *has genus* g *and* k *boundary components, then* $\pi_1(M)$ *has the presentation* (1) *or* (2) *below according as* S *is orientable or not.*

(1) $< a_1, b_1, \cdots, a_g, b_g, c_1, \cdots, c_q, d_1, \cdots, d_k, t :$

$$a_i t a_i^{-1} = t^{\varepsilon_i},\ b_i t b_i^{-1} = t^{\delta_i},\ c_i t c_i^{-1} = t^{\eta_i},$$

$$d_i t d_i^{-1} = t^{\theta_i},\ c_i^{n_i} = t^{s_i},$$

$$c_q = [a_1, b_1] \cdots [a_g, b_g]\, c_1 \cdots c_{q-1} d_1 \cdots d_k >$$

(2) $< a_1, \cdots, a_g, c_1, \cdots, c_q, d_1, \cdots, d_k, t :$

$$a_i t a_i^{-1} = t^{\varepsilon_i},\ c_i t c_i^{-1} = t^{\eta_i},\ d_i t d_i^{-1} = t^{\theta_i},$$

$$c_i^{n_i} = t^{s_i},\ c_q = a_1^2 \cdots a_g^2\, c_1 \cdots c_{q-1} d_1 \cdots d_k > .$$

Furthermore, each $\varepsilon_i, \delta_i, \eta_i, \theta_i$ *is* ± 1 *and* t *is represented by any regular fiber.*

PROOF. The generators, except for t, come from standard generators for a cross section to the fiber bundle $\eta : M^* \to S^*$. We can reduce this cross section to a 2-cell by cutting along arcs "orthogonal" to these generators. Thus we see that M^* is obtained from $B^2 \times S^1$ (the bundle over this 2-cell) by identifying pairs of annuli in its boundary. The relations $a_i t a_i^{-1} = t^{\pm 1}$, etc., come from these identifications, with the exponent -1 occurring when the identification reverses the S^1-direction. The remaining relations come from attaching the solid tori neighborhoods of the singular fibers.

Thus, if S is orientable, then M is orientable if and only if each ε_i, δ_i, η_i, θ_i is $+1$.

Fuchsian Groups

The cyclic group generated by t is normal in $\pi_1(M)$ and is central precisely when each ε_i, δ_i, η_i, θ_i is $+1$. The quotient groups, $\pi_1(M)/\langle t\rangle$, can be presented by adding the relation $t = 1$. These groups are Fuchsian groups which occur also in complex variable theory.

Note that if $\partial M \neq \emptyset$ (i.e. $k > 0$), then the last relation can be eliminated by solving for d_k in terms of the other generators. Thus in this case, $\pi_1(M)/\langle t\rangle$ is just a free product of cyclic groups. Such a group is known to contain a free subgroup of finite index and the following theorem can be adapted to show this; however to simplify notation, we prove

12.2. THEOREM. *Let* $G = \pi_1(M)/\langle t\rangle$ *where* M *is a closed Seifert fibered space as in* 12.1, i.e. *either*

(1) $G = \langle a_1, b_1, \cdots, a_g, b_g, c_1, \cdots, c_q :$

$$c_1^{n_1} = \cdots = c_q^{n_q} = 1,$$

$$c_q = [a_1, b_1][a_g, b_g] c_1 \cdots c_{q-1}\rangle, \quad \text{or}$$

(2) $G = \langle a_1, \cdots, a_g, c_1, \cdots, c_q :$

$$c_1^{n_1} = \cdots = c_q^{n_q} = 1,$$

$$c_q = a_1^2 \cdots a_g^2 c_1 \cdots c_{q-1}\rangle .$$

Assume each $n_i \geq 2$. *Then*

(i) G *contains a subgroup of finite index,* λ, *which is isomorphic to the fundamental group of a closed 2-manifold* T.

(ii) $$\chi(T) = \begin{cases} \lambda\left(2-2g - \sum_{i=1}^{q} \left(1-\frac{1}{n_i}\right)\right) & \text{case (1)} \\ \lambda\left(2-g - \sum_{i=1}^{q} \left(1-\frac{1}{n_i}\right)\right) & \text{case (2)} \end{cases}$$

(iii) G *is finite if and only if:*

Case (1) $g = 0$ *and either* $q \leq 2$ *or* $q = 3$ *and* $\frac{1}{n_1} + \frac{1}{n_2} + \frac{1}{n_3} > 1$.

Case (2) *Either* $g = 1$, $q = 1$, *and* $n_1 = 2$, *or* $g = 0$ *and* $q \leq 2$, *or* $g = 0$ *and* $q = 3$ *and* $\frac{1}{n_1} + \frac{1}{n_2} + \frac{1}{n_3} > 1$.

PROOF. The space, E, obtained from B^2 by identifying each $\ell^{i\theta} \in \partial B^2$ with $\ell^{i\left(\theta+\frac{2\pi}{n}\right)}$ will be called an *n-cap*. We denote by $b(E)$ the image of ∂B^2 under the identification.

Let F be a surface with boundary components $J_1, \cdots, J_q$. For each i, attach an n_i – cap, E_i, to F by identifying $b(E_i)$ with J_i. The resulting space, X, is called a *capped surface*. For appropriate choice of F, we have $\pi_1(X) \cong G$. Let $c(X)$ denote the maximum of $\{n_1, \cdots, n_q\}$.

Let $p : \tilde{X} \to X$ be a finite sheeted cover. Then $\tilde{F} = p^{-1}(F)$ is a connected surface and each component of $p^{-1}(E_i)$ is a union of caps of "order" $\leq n_i$. Let X_1 be obtained from $\tilde{F}$ by adding just one cap from each component of $p^{-1}(E_i)$ for each i. Then X_1 is a capped surface and $c(X_1) \leq c(X)$. Furthermore, $i_* : \pi_1(X_1) \to \pi_1(\tilde{X})$ is an isomorphism, since adding the remaining caps adds redundant relations to $\pi_1(X_1)$.

If the covering $p : \tilde{X} \to X$ is regular, with λ sheets, then each component of $p^{-1}(E_i)$ is a union of q_i (n_i/q_i)-caps and there are λ/q_i components. If $q_i = 1$, then $b(E_i)$ represents an element of the normal subgroup $p_*\pi_1(\tilde{X})$. Thus if G contains a normal subgroup N of finite index which contains none of the elements c_i with $n_i = c(X)$, then we let $\tilde{X}$ be the covering corresponding to N. By the above remarks $c(X_1) < c(X)$ and we can continue by induction to prove part (i).

To obtain N, we note that if X has more than one cap of maximal order, then the corresponding c_i's are mapped nontrivially in the finite group obtained from G by putting the a_i's and b_i's equal to 1 and abelianizing. Thus we may take N to be the kernel of this map. If G contains any proper normal subgroup, N_1, of finite index, then either we can choose $N = N_1$ or the capped surface in the covering of X corresponding to N_1 has two caps of maximal order. In the latter case, the previous argument gives a subgroup N_2 of finite index in N_1 containing none of the elements c_i with $n_i = c(X)$. We obtain N by intersecting N_2 with its finitely many distinct conjugates in G. Thus it suffices to show that G contains a proper normal subgroup of finite index. This is easily established if $g \neq 0$. So suppose $g = 0$. If in addition $q \leq 2$, G is already finite and we may take $N = 1$. So assume $g = 0$ and $q \geq 3$. It is not too difficult to find 2×2 matrices A, B over an appropriately chosen finite field $GF(p^n)$ which satisfy the relations $A^{n_1} = B^{n_2} = (AB)^{n_q} = 1$. Then the map $c_1 \to A$, $c_2 \to B$, $c_q \to AB$, $c_i \to 1$, $i \neq 1, 2, q$ extends to a homomorphism whose kernel is the desired normal subgroup of finite index.

By inducting on $c(X)$, we can obtain a finite sheeted covering $\rho : X^* \to X$ such that for each i each component of $\rho^{-1}(E_i)$ is a union of 1-caps. Thus there is a closed surface $T \subset X^*$ with $i_* : \pi_1(T) \to \pi_1(X^*)$ an isomorphism. Thus $\rho_* \pi_1(T)$ is the desired subgroup giving (i). Let $\lambda = [G : \rho_* \pi_1(T)]$. Now each component of $\rho^{-1}(b(E_i))$ is an n_i-sheeted cover of $b(E_i)$; so $\rho^{-1}(b(E_i))$ has λ / n_i components. Thus T is obtained from $\rho^{-1}(F)$ by adding $\sum_{i=1}^{q} \lambda / n_i$ 2-cells. Hence $\chi(T) = \chi(\rho^{-1}(F)) + \sum_{i=1}^{q} \lambda / n_i$. But $\chi(\rho^{-1}(F)) = \lambda \chi(F)$ and

$$\chi(F) = \begin{cases} 2-2g-q & \text{case (1)} \\ 2-g-q & \text{case (2)} . \end{cases}$$

Putting these together, we get (ii). Finally, G is finite if and only if $\chi(T) > 0$. It is easily checked that this occurs precisely for the cases given in (iii).

Bundles with Periodic Structure Group

Surface bundles over S^1 are always "foliated" by lines and/or circles. In some cases this "foliation" reduces to a Seifert fibration. Specifically let F be a compact surface and $\phi : F \to F$ a homeomorphism. Then the space M_ϕ obtained from $F \times I$ by identifying $(x, 0) \in F \times 0$ with $(\phi(x), 1) \in F \times 1$ is a fiber bundle over S^1 with fiber F. Every such bundle is so obtained. For $x \in F$, the set $\bigcup_{i=-\infty}^{\infty} \phi^i(x) \times I$ becomes a line or a circle in M_ϕ -- a circle precisely if $\phi^k(x) = x$ for some k. If ϕ is periodic, then we have M_ϕ represented as a disjoint union of circles. This decomposition gives M_ϕ the structure of a Seifert fibered space (see 12.6 below).

Note that if $\phi_1, \phi_2 : F \to F$ are isotopic, then an isotopy between 1 and $\phi_2^{-1}\phi_1$ induces a map $F \times I \to F \times I$ which in turn induces a fiber preserving homeomorphism $M_{\phi_1} \to M_{\phi_2}$. It is known (cf. Theorem 6.4 of [21]) that homotopic homeomorphisms of F are isotopic. Two homeomorphisms of F which induce the same automorphism of $\pi_1(F)$, modulo an inner automorphism, are homotopic. Thus M_ϕ is determined by the outer automorphism of $\pi_1(F)$ induced by ϕ.

We can present

$$\pi_1(M_\phi) = \pi_1(F) * \langle t : - \rangle / N$$

where N is the smallest normal subgroup containing

$$\{t\alpha t^{-1}\phi_*(\alpha^{-1}) : \alpha \in \pi_1(F)\}$$

where $\phi_* : \pi_1(F) \to \pi_1(F)$ is induced by ϕ and t is represented by a loop meeting each fiber transversely in a single point. Each element of $\pi_1(M_\phi)$ can be uniquely represented in the form αt^n $\alpha \in \pi_1(F)$. The multiplication is given by

$$(\alpha t^n)(\beta t^m) = \alpha\phi_*^n(\beta)\, t^{n+m} .$$

For future reference, we give some examples.

12.3. EXAMPLES. *The following gives periodic homeomorphisms* ϕ_i *of* $S^1 \times S^1$, *and describes the corresponding manifold* $M_{\phi_i} = M_i$. *Here* S^1 *is represented as the additive group of real numbers modulo* 1.

(1) $\phi_1 = 1$. $M_1 = S^1 \times S^1 \times S^1$.

$\pi_1(M_1) = H_1(M_1) = \langle \alpha, \beta, t : [\alpha, \beta] = [\alpha, t] = [\beta, t] = 1 \rangle$

(2) $\phi_2(x, y) = (-x, -y)$. $\phi_2^2 = 1$.

$\pi_1(M_2) = \langle \alpha, \beta, t : [\alpha, \beta] = 1,\ t\alpha t^{-1} = \alpha^{-1},\ t\beta t^{-1} = \beta^{-1} \rangle$

$H_1(M_2) = Z + Z_2 + Z_2$. ϕ_2 preserves orientation; so M_2 is orientable.

(3) $\phi_3(x, y) = (x, -y)$. $\phi_3^2 = 1$.

$\pi_1(M_3) = \langle \alpha, \beta, t : [\alpha, \beta] = 1,\ t\alpha t^{-1} = \alpha,\ t\beta t^{-1} = \beta^{-1} \rangle$

$H_1(M_3) = Z + Z + Z_2$. ϕ_3 reverses orientation; so M_3 is nonorientable.

(4) $\phi_4(x, y) = (x+y, -y)$. $\phi_4^2 = 1$.

$\pi_1(M_4) = \langle \alpha, \beta, t : [\alpha, \beta] = 1,\ t\alpha t^{-1} = \alpha,\ t\beta t^{-1} = \alpha\beta^{-1} \rangle$

$H_1(M_4) = Z + Z$. M_4 is nonorientable.

(5) $\phi_5(x, y) = (-y, x-y)$, $\phi_5^3 = 1$

$\pi_1(M_5) = \langle \alpha, \beta, t : [\alpha, \beta] = 1,\ t\alpha t^{-1} = \beta,\ t\beta t^{-1} = \alpha^{-1}\beta^{-1} \rangle$

$H_1(M_5) = Z + Z_3$. M_5 is orientable.

(6) $\phi_6(x, y) = (y, -x)$, $\phi_6^4 = 1$,

$\pi_1(M_6) = \langle \alpha, \beta, t : [\alpha, \beta] = 1,\ t\alpha t^{-1} = \beta^{-1},\ t\beta t^{-1} = \alpha \rangle$

$H_1(M_6) = Z + Z_2$. M_6 is orientable.

(7) $\phi_7(x, y) = (-y, x+y)$, $\phi_7^6 = 1$.

$\pi_1(M_7) = \langle \alpha, \beta, t : [\alpha, \beta] = 1,\ t\alpha t^{-1} = \beta,\ t\beta t^{-1} = \alpha^{-1}\beta \rangle$

$H_1(M_7) = Z$. M_7 is orientable.

12.4. LEMMA. *Let* $\phi : S^1 \times S^1 \to S^1 \times S^1$ *be a periodic homeomorphism. Then* M_ϕ *is homeomorphic to one of the seven spaces of 12.3.*

PROOF. ϕ_* is a periodic automorphism of the free abelian group $\pi_1(S^1 \times S^1)$. The minimum polynomial of ϕ_* has degree at most two, has integer coefficients, and distinct p^{th} roots of unity ($p = \text{order } \phi_*$) as roots and therefore must be one of

$$x - 1,\ x + 1,\ x^2 - 1,\ x^2 + 1,\ x^2 - x + 1,\ x^2 + x + 1\ .$$

It is straightforward to check that there is a basis for $\pi_1(S^1 \times S^1)$ with respect to which the matrix of ϕ_* is one of

$$\begin{pmatrix} 1 & 0 \\ 0 & 1 \end{pmatrix}, \begin{pmatrix} -1 & 0 \\ 0 & -1 \end{pmatrix}, \begin{pmatrix} 1 & 0 \\ 0 & -1 \end{pmatrix}, \begin{pmatrix} 1 & 0 \\ 1 & -1 \end{pmatrix},$$

$$\begin{pmatrix} 0 & 1 \\ -1 & -1 \end{pmatrix}, \begin{pmatrix} 0 & -1 \\ 1 & 0 \end{pmatrix}, \begin{pmatrix} 0 & 1 \\ -1 & 1 \end{pmatrix}.$$

For example, suppose the minimum polynomial is $x^2 - x + 1$. Of all bases, choose one which minimizes $|q_{11}|$ in the matrix (q_{ij}) of ϕ_*. Note that $q_{11} + q_{22} = 1$ and $q_{11} q_{22} - q_{12} q_{21} = 1$. We must have $q_{11} = 0$. If not, then one of $|q_{12}| \leq |q_{11}|$, $|q_{21}| \leq |q_{11}|$ must hold; otherwise $q_{11}^2 - q_{11} + 1 = |q_{11}^2 - q_{11} + 1| = |q_{12} q_{21}| \geq (|q_{11}| + 1)^2$ which is impossible if $q_{11} \neq 0$. Since conjugation by $\begin{pmatrix} 1 & 0 \\ \pm 1 & 1 \end{pmatrix}$ (or by $\begin{pmatrix} 1 & \pm 1 \\ 0 & 1 \end{pmatrix}$) replaces q_{11} by $q_{11} \pm q_{21}$ (or by $q_{11} \pm q_{12}$), we can reduce q_{11}. Thus $q_{11} = 0$; hence $q_{22} = 1$ and $q_{12} q_{21} = -1$ and the reduction follows.

Then $\phi_* = \phi_{i*}$ for ϕ_i one of the maps defined in 12.3 and the conclusion follows.

12.5. EXERCISE. *Carry out* 12.4 *for periodic homeomorphisms of the Klein bottle* (refer to 2.12).

12.6.. LEMMA. *Let* M *be a fiber bundle over* S^1 *with fiber a compact surface* F; *so* $M = M_\phi$ *for some homeomorphism* $\phi : F \to F$. *If* $\pi_1(M)$

contains an infinite cyclic, normal subgroup, N, *with* $N \not\subset \pi_1(F)$, *then* N *is central in* $\pi_1(M)$ *and for some* $k > 0$, ϕ_*^k *is an inner automorphism of* $\pi_1(F)$. *If* M *is orientable, then* M *is a Seifert fibered space with orientable Seifert surface and* N *lies in the subgroup generated by any regular fiber.*

PROOF. Let αt^k, $\alpha \in \pi_1(F)$, generate N. Since $N \not\subset \pi_1(F)$, $k \neq 0$. Since N is normal, for any $\beta \in \pi_1(F)$ and any m either

$$(\beta t^m)(\alpha t^k)(\beta t^m)^{-1} = \alpha t^k$$

or

$$(\beta t^m)(\alpha t^k)(\beta t^m)^{-1} = (\alpha t^k)^{-1} .$$

The second case, with $m = 0$, gives $\beta\alpha\phi_*^k(\beta^{-1})t^k = \phi_*^{-k}(\alpha)t^{-k}$. This is impossible since $t^{2k} \notin \pi_1(F)$. Thus the first holds for all β, m and thus N lies in the center of $\pi_1(M)$. Furthermore, for every $\beta \in \pi_1(F)$ $\beta\alpha\phi_*^k(\beta^{-1}) = \alpha$. Thus ϕ_*^k is an inner automorphism (conjugation by α^{-1}).

Now suppose M is orientable. Then F (being 2-sided) is orientable and $\phi: F \to F$ is orientation preserving. By a theorem of J. Nielsen [79], ϕ is homotopic, hence isotopic, to a periodic homeomorphism $\phi_1 : F \to F$. Thus $M = M_{\phi_1}$. By our earlier remarks, M is a disjoint union of circles. If C is one of these, then $C \cap F$ is a finite set which is the complete orbit, under the action of ϕ_1, of any one of its points. Let D be a 2-cell neighborhood of one of the points of $C \cap F$ such that $D, \phi_1(D), \cdots, \phi_1^{m-1}(D)$ are pairwise disjoint and $\phi_1^m(D) = D$. By another theorem of Nielsen [78], $\phi_1^m | D$ is equivalent to a rotation about its fixed point $D \cap C$ whose period is n/m ($n = $ order ϕ_1). From this it is easy to construct a fibered solid torus neighborhood of C. Thus M is Seifert fibered. Observe that a fiber, C, is regular if and only if $\#(C \cap F) = n$ (= order ϕ_1).

Thus a regular element represents an element of the form γt^n and this element is central in $\pi_1(M)$. If the center of $\pi_1(M)$ is cyclic, it must be generated by γt^n; otherwise we have a central element of the form δt^r where $r|n$. The previous computations show that ϕ_{1*}^r is an inner automorphism and thus ϕ_1 is homotopic to a homeomorphism of period r. This justifies our claim and proves, in this case, that N is in the subgroup generated by a regular fiber. If the center $z(\pi_1(M))$ is not cyclic, then $z(\pi_1(M)) \cap \pi_1(F) \neq 1$; so F is a torus or an annulus. If F is an annulus, then $M = S^1 \times S^1 \times I$ (not the twisted I-bundle over a Klein bottle – since its fundamental group has cyclic center) and a (possibly different) Seifert fibration can be chosen so that N lies in the group generated by the fiber. If F is a torus, we refer to 12.3. Note that an element of $\pi_1(F) \cap z(\pi_1(M))$ must be fixed by ϕ_{i*}. One checks that among the orientable cases (1, 2, 5, 6, 7 of 12.3), only ϕ_{1*} leaves a nontrivial element fixed. Thus $M = S^1 \times S^1 \times S^1$ and the fibration can be chosen so that N lies in the group of the fiber.

Cyclic Normal Subgroups

A 3-manifold M is said to be *sufficiently large* if M contains a properly embedded, 2-sided, incompressible surface. Recall that if $|H_1(M)| = \infty$ (and M is compact), then by 6.6, M is sufficiently large – in fact M contains a nonseparating incompressible surface. Note also that by 6.7, $|H_1(M)| = \infty$ unless M is orientable and ∂M contains at most 2-spheres or M is nonorientable and ∂M contains a projective plane. There are, however, closed, orientable, sufficient large 3-manifolds with finite (trivial) first homology. Examples can be obtained by appropriately sewing together two knot spaces along their boundaries.

The next two results describe the structure of a compact, sufficiently large 3-manifold, M, whose fundamental group contains an infinite cyclic, normal subgroup. Theorem 12.7 treats the case $|H_1(M)| = \infty$. The description is a bit awkward, but it reduces (12.8) in the orientable case to saying that M is Seifert fibered. The case $|H_1(M)| < \infty$ is also treated in 12.8. It is conceivable that M is a "generalized Seifert fibered space" in all cases, but there are some technical obstacles to proving this – see the remarks following 12.8.

These results extend the work of Waldhausen [103].

12.7. THEOREM. *Let* M *be a compact 3-manifold with* $M = \mathcal{P}(M)$. *If* M *is closed and orientable, assume* $H_1(M)$ *is infinite. If* $\pi_1(M)$ *contains an infinite cyclic, normal subgroup,* N, *then either* M *is homotopy equivalent to* $P^2 \times S^1$ *or there is a sequence* $M = M_0 \supset M_1 \supset \cdots \supset M_k$ *where*

(i) M_{i+1} *is obtained from* M_i *by cutting open along a properly embedded, nonseparating, 2-sided, incompressible surface* F_i *in* M_i *where* $\chi(F_i) = 0$,

(ii) $N < i_*\pi_1(F_i)$ *for all* i, *and*

(iii) M_k *is a fiber bundle over* S^1 *with fiber a compact surface* $F: M_k = F \times I/(x,0) = (\phi(x),1)$ *where some power of* $\phi_*: \pi_1(F) \to \pi_1(F)$ *is an inner automorphism.*

PROOF. If M contains a P^2, then $\pi_1(M)$ contains a subgroup isomorphic to $Z + Z_2$. By 9.12 and the fact that $\pi_1(M)$ is not a free product (11.2), $\pi_1(M) \cong Z + Z_2$ and M is homotopy equivalent to $P^2 \times S^1$ (11.6). If M contains an essential 2-sphere, then M is a 2-sphere bundle over S^1. So we may assume M is P^2-irreducible.

We must have $H_1(M)$ infinite. This is hypothesis if M is closed and orientable and follows from 6.7 in the other cases.

Thus by 6.6, M contains a properly embedded, nonseparating, 2-sided, incompressible surface F. If M is orientable and $\partial M \neq \emptyset$, we may assume that $\partial F \neq \emptyset$ (6.8).

Case 1. $N \not< \pi_1(F)$. Let R_0 denote M cut open along F. Then we have copies F'_0, F''_0 of F in ∂R_0 and a map $g_0: R_0 \to M$ such that g_0 maps $R_0 - (F'_0 \cup F''_0)$ homeomorphically to $M - F$ and maps each of F'_0, F''_0 homeomorphically to F. For each integer $n\,(-\infty < n < \infty)$, take a copy R_n of R_0 with corresponding copies F'_n, F''_n of the surfaces F'_0, F''_0 and g_n of the map g_0. For each n identify F''_{n-1} with F'_n by identifying $x \in F''_n$ with $(g_n|F'_n)^{-1}(g_{n-1}(x))$. Call the resulting space $\tilde{M}$ and denote the common image of F''_{n-1} and F'_n by F_n. The maps

$g_n : R_n \to M$ induce a covering map $p : \tilde{M} \to M$. This covering is regular. The group of covering translations is infinite cyclic with a generator τ satisfying $\tau(R_n) = R_{n+1}$ and $\tau(F_n) = F_{n+1}$. Thus $\pi_1(M)$ is an extension of $p_*\pi_1(\tilde{M})$ by an infinite cyclic group whose generator, t, (corresponding to τ) is represented by a loop which crosses F transversely in a single point x_0. Any lifting of this loop is a path in R_n from a point $\tilde{x}_{n-1} \in F_{n-1}$ to $\tilde{x}_n \in F_n$.

We wish to show that R_0 is homeomorphic to $F_0 \times I$ (with F_0 corresponding to $F_0 \times 0$ and F_1 to $F_0 \times 1$). Suppose this is not the case, then by 10.2, none of the maps

$$\pi_1(F_n) \to \pi_1(R_n) \quad \text{or} \quad \pi_1(F_n) \to \pi_1(R_{n-1})$$

is epic. Thus $\pi_1(F_0)$ has infinite index in $\pi_1(\tilde{M})$. We have $\pi_1(\tilde{M}) = \pi_1(\tilde{M}_-) \underset{\pi_1(F_0)}{*} \pi_1(\tilde{M}_+)$ where $\tilde{M}_- = \bigcup_{n<0} R_n$ and $\tilde{M}_+ = \bigcup_{n\geq 0} R_n$. It is not difficult to show, using normal form for elements, that a cyclic normal subgroup of an amalgamated free product either lies in the amalgamating subgroup or the amalgamating subgroup has index two in each factor. Thus, since $\pi_1(F_0)$ has infinite index in $\pi_1(\tilde{M})$ and $N \not< \pi_1(F)$, we must have $N \not< p_*\pi_1(\tilde{M})$.

Let yt^r generate N where $y \in p_*\pi_1(\tilde{M})$. By the above, $r \neq 0$. Now N is normal; so for any $z \in p_*\pi_1(\tilde{M})$ and any s, either

$$(zt^s)(yt^r)(zt^s)^{-1} = yt^r$$

or

$$(zt^s)(yt^r)(zt^s)^{-1} = (yt^r)^{-1} .$$

Since $p_*\pi_1(\tilde{M})$ is normal in $\pi_1(M)$, the second possibility would give $t^{2r} \in p_*\pi_1(\tilde{M})$ which is impossible. Thus N is, in fact, central in $\pi_1(\tilde{M})$. Now for any m, $t^{-m}(yt^r)t^m = (t^{-m}yt^m)t^r \in N$ and $y_1 = t^{-m}yt^m \in p_*\pi_1(\tilde{M})$. By choosing m large enough, there is a loop α in $\tilde{M}_-$ based at $\tilde{x}_0$ such that $p_*([\alpha]) = y_1$. Let β be any loop in R_0 based at $\tilde{x}_0$. Let γ be a path in $R_0 \cup \cdots \cup R_{r-1}$ from $\tilde{x}_0$ to $\tilde{x}_r$ such that $[p \circ \gamma] = t^r$, and consider the loop $\theta = \beta^{-1} \cdot (\alpha\gamma) \cdot (\tau^r \circ \beta) \cdot (\alpha\gamma)^{-1}$. Now $p_*([\theta]) = [p_*[\beta]^{-1}, y_1 t^r]$. But $y_1 t^r$ is central; so $[\theta] = 1$. Using the incompressibility of each F_n, the fact that θ meets R_r only in the loop $\tau^r \circ \beta$, and the fact that θ is null homotopic in $\tilde{M}$, we see that $\tau^r \circ \beta$ is homotopic, in R_r, to a loop in F_r. Since β was arbitrary, we have shown that $\pi_1(F_r) \to \pi_1(R_r)$ is epic. Thus $\pi_1(F_0) \to \pi_1(R_0)$ is epic, R_0 is a product $F_0 \times I$, and so M is a fiber bundle over S^1 with fiber F. This completes the proof in Case 1. The last part of the conclusion (that some power of ϕ_* is inner) follows from 12.6.

Case 2. $N < \pi_1(F)$. Since $\pi_1(F)$ contains an infinite cyclic normal subgroup, we must have $\chi(F) = 0$. Let $F_0 = F$ and let M_1 be M cut open along F_0. Since $\partial M_1 \neq \emptyset$, M_1 contains a properly embedded, nonseparating, 2-sided incompressible surface F_1. Now $N < \pi_1(M_1)$. If $N \not< \pi_1(F_1)$, Case 1 applied to M_1 completes the proof. If $N < \pi_1(F_1)$, we let M_2 be M_1 cut open along F_1 and continue. The proof is completed by showing that we cannot continue indefinitely without reducing to Case 1. If, at some stage, r of the F_i's already constructed are closed (if M is orientable, we need at most one closed F_i), then these F_i's are pairwise disjoint and their union doesn't separate M. Thus we can retract M to a graph dual to these F_i's. This retraction induces an epimorphism of $\pi_1(M)$ to a free group of rank r. Thus, $r \leq p\pi_1(M)$. Hence there is some integer, m, such that for $i > m$, $\partial F_i \neq \emptyset$. We claim that we may choose $F_{m+1}, F_{m+2}, \cdots$ to be pairwise disjoint. For suppose we have succeeded in doing thus up to F_{m+k}. Let J be a component of ∂F_{m+k+1} and T the component ∂M_{m+k} containing J. Now T is a torus or a Klein bottle (since $\pi_1(T) \cap N \neq 1$) and J is 2-sided in T. We have a collection $A_1, \cdots, A_s$ of pairwise disjoint annuli and/or Möbius bands in T corresponding to the copies of $F_{m+1}, \cdots, F_{m+k}$ lying in T. If T is compressible in M_{m+k}, then M_{m+k} is a solid torus or a solid Klein bottle and we stop at this stage. If T is incompressible, then some power of J is homotopic in M_k to some power of a core of (any) A_i in

M_k and hence homotopic to this core in T. If some A_i is an annulus, then J is isotopic to a core of A_i. If some A_i is a Möbius band, then J is isotopic to ∂A_i. Thus we can adjust F_{m+k+1} to be disjoint from $F_{m+1}, \cdots, F_{m+k}$. It now follows that we can continue at most $p(\pi_1(M_m))$ steps beyond M_m; since we can retract M_m onto a graph dual to $F_{m+1}, F_{m+2}, \cdots$.

12.8. COROLLARY. *Let* $M = \mathcal{P}(M)$ *be a compact, orientable, sufficiently large 3-manifold and let* N *be an infinite cyclic normal subgroup of* $\pi_1(M)$. *Then either*

(i) M *is a Seifert fibered space and* N *lies in the group generated by any regular fiber, or*

(ii) $M = M_1 \cup M_2$ *with* $M_1 \cap M_2 = \partial M_1 = \partial M_2$ *where either* M_i *is a twisted* I*-bundle over a closed nonorientable surface or* $\pi_1(M_i) = Z_2$.

If $|H_1(M)| = \infty$ *or if* N *is central in* $\pi_1(M)$, *then Case* (i) *holds.*

PROOF. Let F be a properly embedded, 2-sided, incompressible surface in M. If F doesn't separate M, then 12.7 applies to give the sequence

$$M = M_0 \supset M_1 \supset \cdots \supset M_k$$

where M_k is a fiber bundle over S^1. Since $N < \pi_1(M_k)$, 12.6 gives a Seifert fibration of M_k with N in the group generated by a regular fiber. We have a pair A_1, A_2 of annuli or tori in ∂M_k which are identified to yield M_{k-1}. If M_k is a solid torus, then a fibration of $A_1 \cup A_2$ can be extended to one of M_k (possibly with a singular fiber). In any other case, ∂M_k is incompressible and since $N < \pi_1(A_i)$, it follows that we can modify the fibration of M_k by an isotopy to obtain one which extends to a Seifert fibration of M_{k-1}. Continuing in this way, we obtain the desired conclusion.

Now suppose that F separates M, and denote the closures of the components of M–F by M_1 and M_2.

If $N < \pi_1(F)$ (which must be the case if N is central in $\pi_1(M)$), then the preceding case gives us Seifert fibrations of M_1 and M_2. Since F is incompressible, the two fibrations must be compatible (up to isotopy) on F and we obtain the desired Seifert fibration of M.

If $N \not< \pi_1(F)$, then $\pi_1(F)$ has index two in each of $\pi_1(M_1), \pi_1(M_2)$. If $F = S^2$, the best we can say is that M_i is homotopy equivalent to P^2. In any other case, 10.5 applies to show that M_i is a twisted I-bundle over a closed nonorientable surface.

12.9. REMARKS. It seems likely that in 12.8 case (ii) (except for the cases $\pi_1(M_i) = Z_2$ which involve the Poincaré conjecture), M is also Seifert fibered. For M_i is the mapping cylinder of a double covering projection $p_i : F \to F_i$. The fiber of M_i which begins at $x \in F$ ends at $\tau_i(x)$ where $\tau_i : F \to F$ is the corresponding covering translation. Thus we may begin at $x \in F$, proceed along a fiber of M_1 to $\tau_1(x)$, then along a fiber of M_2 to $\tau_2\tau_1(x)$, etc. If $\tau_2\tau_1$ is periodic, then it easily follows that M is Seifert fibered. It does follow that $(\tau_2\tau_1)_* : \pi_1(F) \to \pi_1(F)$ induces a periodic outer automorphism. This can be seen by a direct computation or by noting that if $p : \tilde{M} \to M$ is the double covering with N central in $p_*\pi_1(\tilde{M})$, then $\tilde{M}$ is a bundle over S^1 ($p^{-1}(M_i)$ is a product. Then $\tilde{M} = M_\phi$ where ϕ is isotopic to $\tau_2\tau_1$. By Nielsen's theorem [79], ϕ is isotopic to a periodic homeomorphism $\phi_1 : F \to F$ and as in 12.6, $\tilde{M} = M_{\phi_1}$ is Seifert fibered. The problem is in applying Nielsen's theorem "equivariantly" with respect to the covering translation of $\tilde{M}$.

Another approach to showing that M is Seifert fibered is to show that $\pi_1(M)/N$ is a Fuchsian group and proceed as in the proofs of 11.6 and 11.10 by doing "surgery" on a map of M to a capped surface. Now $\pi_1(M)/N$ contains the Fuchsian group $\pi_1(\tilde{M})/N$ as a subgroup of index two, and Theorem 1′ of [115] may apply to show that $\pi_1(m)/N$ is Fuchsian. However, there may be some question about this result. (see footnote to [115], also [116]); so we do not pursue this approach.

It also seems likely, that by generalizing the definition of a Seifert fibered space, Theorem 12.7 would lead to a result analogous to 12.8 in the nonorientable ease. One would include the possibility that a fiber have a fibered solid Klein bottle neighborhood in generalizing the definition. This would have the disadvantage of allowing infinitely many singular fibers: in fibering a solid Klein bottle (mobius band $\times$ interval), one has an annulus of singular fibers. One would need an extension of Nielsen's theorem on periodic homeomorphisms to nonorientable surfaces to extend 12.8 to nonorientable manifolds.

Centers

We can now describe the center $z(\pi_1(M))$ of the fundamental group of a compact 3-manifold M. Note that if the center is finite ($\neq 1$), then by 9.8 and 9.12, $|\pi_1(M)| < \infty$. If $z(\pi_1(M)) = Z$ and M is sufficiently large, then 12.7 and/or 12.8 apply. There are only a few other cases if $z(\pi_1(M))$ is finitely generated (it is not known whether this is always the case). For this we do not need the assumption that M be sufficiently large.

12.10. THEOREM. *Let* $M = \mathcal{P}(M)$ *be a compact 3-manifold with* $|\pi_1(M)| = \infty$. *If the center,* $z(\pi_1(M))$ *of* $\pi_1(M)$ *is nontrivial, finitely generated, and not cyclic, then either*

(i) M *is an I-bundle over a torus,*

(ii) M *is one of the examples* M_1, M_3, *or* M_4 *of 12.3, or*

(iii) M *is homotopy equivalent to* $P^2 \times S^1$.

PROOF. Since $\pi_1(M)$ is not a free product, it follows from 9.8 and 9.12 that if $\pi_1(M)$ has torsion, then $\pi_1(M) \cong Z + Z_2$ giving (iii).

If $\pi_1(M)$ is torsion free, then $z(\pi_1(M))$ must contain a (normal) subgroup, K, isomorphic to $Z + Z$. If K has finite index in $\pi_1(M)$, M must be an I-bundle over a torus (10.6).

If K has infinite index in $\pi_1(M)$, then by 11.1, either M is a bundle over S^1 with fiber a surface F or $M = M_1 \cup M_2$ where $M_1 \cap M_2 = \partial M_1 =$

$\partial M_2 = F$ and M_i is a twisted I-bundle. In either case $K < \pi_1(F)$ (in fact, $K < z(\pi_1(F))$); so F is a torus. In the first case, putting $M = M_\phi$ for a homeomorphism $\phi: F \to F$, we see that $\phi_*|K$ is the identity. Thus ϕ_* is the identity and $M = S^1 \times S^1 \times S^1$. In the second case, M_i is a twisted I-bundle over a torus (not a Klein bottle), since K is central in $\pi_1(M_i)$. Thus $\pi_1(F)$ is central in $\pi_1(M)$. Then M is doubly covered by a torus bundle over S^1 and F lifts homeomorphically to a fiber. By the first case, this double cover is $S^1 \times S^1 \times S^1$. The next theorem shows that M also has the structure of a bundle over S^1 with fiber a torus F_1. We do not have $K < \pi_1(F_1)$; so by 12.6, the attaching map $\phi: F_1 \to F_1$ must induce a periodic isomorphism of $\pi_1(F_1)$. Thus M is among the examples given in 12.3. Now $K \cap \pi_1(F_1) \neq 1$; so some element of $\pi_1(F_1)$ is fixed by ϕ_*. One checks the examples to see that only (1), (3), and (4) have this property. This completes the proof. One should observe that these manifolds have many different descriptions; for example (3) of 12.3 is (Klein bottle) $\times S^1$ and also is the union of two twisted I-bundles over a torus where the fundamental group of their common boundary is generated by α and t^2.

Cyclic Actions on $S^1 \times S^1 \times S^1$

We show that the quotient of $S^1 \times S^1 \times S^1$ by a free cyclic action is one of the torus bundles over S^1 given in 12.3.

12.11. THEOREM [40]. *Let* $M = \mathcal{P}(M)$ *be a compact* 3-*manifold for which there is an exact sequence*

$$1 \to A \to \pi_1(M) \overset{\eta}{\to} Z_p \to 1$$

where $A \cong Z + Z + Z$. *Then* M *is one of the torus bundles over* S^1 *described in* 12.3. *Furthermore, there is a cyclic covering* $p: S^1 \times S^1 \times S^1 \to M$ *with either* 1, 2, 3, 4, *or* 6 *sheets and with* $A < p_*\pi_1(S^1 \times S^1 \times S^1)$.

PROOF. There is a maximal abelian, normal subgroup A_1 containing A. Since A has finite index in A_1 and $\pi_1(M)$ is torsion free (9.8 and 9.12), $A_1 \cong Z + Z + Z$. We have an exact sequence

$$1 \longrightarrow A_1 \longrightarrow \pi_1(M) \xrightarrow{\eta_1} Z_{p_1} \longrightarrow 1$$

where p_1 divides p.

Choose $z \in \pi_1(M)$ such that $\eta_1(z)$ generates Z_{p_1}. Then we have an automorphism $\psi : A_1 \to A_1$ given by $\psi(a) = z a z^{-1}$.

Now $z^{p_1} \in A_1$, so ψ is periodic of order q a divisor of p_1. In fact $q = p_1$; otherwise the group generated by A_1 together with z^q is a larger abelian normal subgroup.

If $p_1 = 1$, $\pi_1(M) = A$, and $M = S^1 \times S^1 \times S^1$; so assume $p_1 > 1$.

Choose an element $x_1 \in A_1$ which is not a proper power and such that $z^{p_1} = x_1^n$ for some $n \neq 0$. Now $\psi(x_1^n) = zx_1^n z^{-1} = zz^{p_1}z^{-1} = x_1^n$; so $\psi(x_1) = x_1$ (i.e., $[z, x_1] = 1$). Note that $d = (n, p_1) = 1$; otherwise $(x_1^{n/d} z^{-p_1/d})^d = 1$ contrary to the fact that $\pi_1(M)$ is torsion free. Also by replacing z by $x_1^m z^{\pm 1}$, for some m, we may assume that $1 \leq n \leq [\frac{1}{2} p_1]$.

We switch to additive notation for A_1. The minimal polynomial, m(x), of ψ divides $x^{p_1} - 1$ and therefore has integer coefficients and distinct roots of unity as roots. Now 1 is a characteristic root of m(x), since $\psi(x_1) = x_1$; so as in 12.4 $m(x) = (x-1) f(x)$ where either

(i) $f(x) = x + 1$ (iii) $f(x) = x^2 + x + 1$

(ii) $f(x) = x^2 + 1$ (iv) $f(x) = x^2 - x + 1$.

Now p_1 is the smallest positive integer for which $x^{p_1} - 1$ is divisible by m(x). Direct computation yields

$$p_1 = \begin{cases} 2; & \text{case (i)} \\ 4; & \text{case (ii)} \\ 3; & \text{case (iii)} \\ 6; & \text{case (iv)} . \end{cases}$$

Let $B = \{a \epsilon A_1 : (\psi-1)(a)=0\}$ and $C = \{a \epsilon A_1 : f(\psi)(a)=0\}$. Then by standard arguments $B \cap C = \{0\}$, $\psi(B) = B$ and $\psi(C) = C$ (field coefficients are not needed for this). We cannot conclude that $A_1 = B \oplus C$; however, $k = g(x)(x-1) + f(x)$ for some $g(x) \epsilon Z[x]$ and $k = f(1) \leq 3$. Thus $kA_1 < B \oplus C$; so $B \oplus C$ has finite index in A_1. Moreover, $A_1 = B \oplus C$ if and only if $f(\psi)(A_1) = f(\psi)(B)$. For if $A_1 = B \oplus C$, then $f(\psi)(A_1) = f(\psi)(B) \oplus f(\psi)(C) = f(\psi)(B)$. Conversely, if $f(\psi)(A_1) = f(\psi)(B)$, then for $a \epsilon A_1$, there is some $b \epsilon B$ with $f(\psi)(a) = f(\psi)(b)$. Then $a = b + (a-b) \epsilon B \oplus C$. Now $f(\psi)(B) = kB$, $(k = f(1))$ and $B < f(\psi)(A_1) < kB$. Thus if $f(1) = 1$, then $A_1 = B \oplus C$. This happens in case (iv) $(f(x) = x^2 - x+1)$.

Since B and C are invariant under ψ, they are normal in $\pi_1(M)$; in fact B is central in $\pi_1(M)$. Also A_1/B and A_1/C are torsion free; so each of B, C is a direct summand of A_1.

Now x_1 is indivisible in A_1; so we can extend to a basis $\{x_1,x_2,x_3\}$ for A_1 where $\{x_1,\cdots,x_r\}$ (r = rank B) is a basis for B and, in case $A_1 = B \oplus C$, $\{x_{r+1},\cdots,x_3\}$ is a basis for C.

Now we have a presentation (in multiplicative notation):

$$\pi_1(M) = \langle x_1,x_2,x_3,z : [x_i,x_j] = 1,\ zx_iz^{-1} = \psi(x_i),\ z^{p_1} = x_1^n \rangle .$$

Recall that $(n,p_1) = 1$ and $n \leq [\frac{1}{2} p_1]$. Since $p_1 = 2,3,4,$ or 6, $n = 1$.

Suppose that the basis can be chosen so that the group, D, generated by x_2 and x_3 is normal in $\pi_1(M)$. This will certainly be the case if $A_1 = B \oplus C$. Then $\pi_1(M)/D = \langle x_1,z : zx_1z^{-1} = x_1,\ z^{p_1} = x_1 \rangle$ is infinite cyclic. By 11.6, M is a fiber bundle over S^1 whose fiber, F, satisfies $\pi_1(F) = D$. Thus F is a torus. Furthermore, $x_1 \not\epsilon \pi_1(F)$ is a central element. By 12.6, the map $\phi : F \to F$ describing $M = M_\phi$ induces a periodic automorphism of $\pi_1(F)$. The conclusion then follows from 12.4.

We now suppose that $f(\psi)(A_1) \neq kB$ (and $f(x)$ is one cases (i), (ii), or (iii). The proof is complete in all other cases.

If rank $B = 1$, then $f(\psi)(A_1) = sB$ for some divisor, s, of k. Since $k \leq 3$, $s = 1$. Thus there is some $y \in A_1$ with $f(\psi)(y) = x_1$. In cases (i) and (iii) $f(x) = x^{p_1-1} + x^{p_1-2} + \cdots + 1$. Thus $\psi^{p_1-1}(y) + \psi^{p_1-2}(y) + \cdots + y = x_1$. Written multiplicatively:

$$(z^{p_1-1}yz^{1-p_1})(z^{p_1-2}yz^{2-p_1})\cdots y = x_1$$

giving

$$z^{p_1}(z^{-1}y)^{p_1} = x_1 .$$

Since $z^{p_1} = x_1$, $(z^{-1}y)^{p_1} = 1$. This is impossible since $\pi_1(M)$ is torsion free. In case (ii), $f(x) = x^2 + 1$ and $p_1 = 4$. Thus

$$z^2yz^{-2}y = x_1 ,$$

$$z^4(z^{-2}y)^2 = x_1 ,$$

$$(z^{-2}y)^2 = 1 ,$$

which is a contradiction.

If rank $B = 2$, then $f(x) = x+1$, $p_1 = 2$, and $\psi(x_3) = \alpha x_1 + \beta x_2 - x_3$ for some $\alpha, \beta \in Z$. By replacing x_3 by $x_3 + \gamma x_1 + \delta x_2$ for appropriate γ, δ, we may assume $\alpha, \beta = 0, 1$. If $\alpha = 0$, then $D = gp(x_2, x_3)$ is normal in $\pi_1(M)$. If $\alpha = \beta = 1$, then, putting $x_2' = x_1 + x_2$, we have $D = gp(x_2', x_3)$ normal. In either case the proof follows as above. If $\alpha = 1$, $\beta = 0$, then we have the contradiction that $(z^{-1}x_3)^2 = 1$. This completes the proof.

Note that in the above proof, the fact that $\pi_1(M)$ is torsion free forced the conclusion that $A_1 = B \oplus C$ except possibly in the case rank $B = 2$. It need not follow in this case that $A_1 = B \oplus C$; for the action $\psi(x_1) = x_1$, $\psi(x_2) = x_2$, $\psi(x_3) = x_2 - x_3$ gives example (4) of 12.3. In this case, $A_1/(B \oplus C) = Z_2$ and $\pi_1(M)/A = Z_2 * Z_2$. This reflects the fact that M is also the union of two twisted I-bundles over a Klein bottle.

CHAPTER 13

CLASSIFICATION OF P^2-IRREDUCIBLE, SUFFICIENTLY LARGE 3-MANIFOLDS

In this chapter we present the important results of Waldhausen [105] (as extended in [22] and [36]) referred to in the title.

For this, let $\mathfrak{W}$ be the class of all compact 3-manifolds which are

(i) P^2-irreducible and

(ii) sufficiently large – i.e. contain a properly embedded, 2-sided, incompressible surface.

While it is known (see 13.11 below) that for $M, N \in \mathfrak{W}$ an (abstract) isomorphism between $\pi_1(M)$ and $\pi_1(N)$ need not imply topological equivalence of M and N, we show (13.9) that if there is an isomorphism $\phi : \pi_1(M) \to \pi_1(N)$ which is compatible with the inclusions of the fundamental groups of boundary components, then there is indeed a homeomorphism $h : M \to N$ which induces ϕ. The condition on ϕ, which we define precisely later, will be abbreviated by saying that ϕ preserves the peripheral structure.

This follows from the more general result (13.6) that a map $f : (M, \partial M) \to (N, \partial N)$ which induces a monomorphism of fundamental group systems is, except for some obvious exceptions, homotopic to a covering map.

We consider only P^2-irreducible 3-manifolds to avoid the Poincaré conjecture and to avoid some difficulties involved with connected sums. See the comments following 13.10.

It is clear, by considering lens spaces, that some additional structure must be imposed to obtain results of this sort. While the condition of "being sufficiently large" may not be the weakest possible condition, it is a highly convenient one for the following reasons. If M is sufficiently large and we cut M along an incompressible surface $F(\neq S^2, P^2)$ to obtain a manifold M_1, then by 6.6 and 6.7, M_1 is again sufficiently large. Thus we can cut M_1 along an incompressible surface F_1 to obtain a manifold M_2, and by continuing in this way, generate a sequence $M \supset M_1 \supset M_2 \supset \cdots$. It appears that the manifolds are getting more complicated – e.g. the genus of the boundary goes up if the surface along which we cut has negative euler characteristic. However, we prove in 13.3 that this process must terminate with some M_n all of whose components are 3-cells. This means that the M_j are getting closer to being cubes with handles; so that the cuts will eventually be along 2-cells. This is analogous to the well-known procedure for reducing a 2-manifold to a 2-cell by cuts along a simple closed curve (if closed) and then along a sequence of arcs. It has the same advantage of dimension reduction: the study of M is reduced to the study of (more delicate) properties of the surfaces along which the pieces of M are reattached.

The Analogue for Surfaces

We begin with a 2-dimensional analogue of Theorem 13.6. The proof we outline is similar to the proof we give for 13.6.

13.1 THEOREM. *Suppose* F *and* G *are compact 2-manifolds with* $\pi_1(F) \neq 1$. *Let* $f:(F, \partial F) \to (G, \partial G)$ *be a map such that* $f_*: \pi_1(F) \to \pi_1(G)$ *is monic. Then there is a homotopy* $f_t:(F, \partial F) \to (G, \partial G)$ *with* $f_0 = f$ *and either*

(i) $f_1: F \to G$ *is a covering map, or*

(ii) F *is an annulus or Möbius band and* $f_1(F) \subset \partial G$.

If for some component J *of* ∂F, $f|J: J \to f(J)$ *is a covering map, we can require* $f_t|J = f|J$ *for all* t.

PROOF. First, assume $\partial G \neq \emptyset$. Since f_* is monic, $\partial F \neq \emptyset$. For each component J of ∂F, there is a component K of ∂G with $f(J) \subset K$. The maps $\pi_1(J) \to \pi_1(F)$, $\pi_1(K) \to \pi_1(G)$ are monic; so $(f|J)_*: \pi_1(J) \to \pi_1(K)$

is monic. Thus, after modifying by a homotopy, we may assume $f|J : J \to K$ is a covering map. We do this for each component of ∂F.

Now consider the commutative diagram

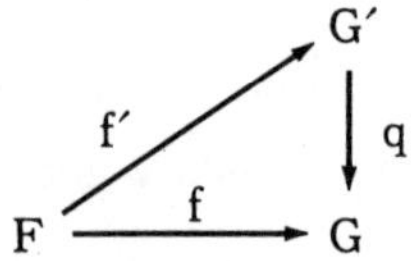

where q is a covering map and $f'_* : \pi_1(F) \to \pi_1(G')$ is an isomorphism. Note that f' restricts to a covering map on each component of ∂F.

Suppose that $f'|\partial F$ is not an embedding. We will show that this case leads to conclusion (ii). With this assumption, there is a path $\alpha : (I, \partial I) \to (F, \partial F)$ satisfying

(*) $\alpha(0) \neq \alpha(1)$, $f'(\alpha(0)) = f'(\alpha(1))$, and the loop $f' \circ \alpha$ is nullhomotopic in G'.

To see this, let $\gamma : (I, \partial I) \to (F, \partial F)$ be any path with $\gamma(0) \neq \gamma(1)$ and $f'(\gamma(0)) = f'(\gamma(1))$. Since f'_* is epic, there is a loop β based at $\gamma(0)$ such that $f_*[\beta] = [f \circ \gamma]^{-1} \in \pi_1(G')$. Then $\alpha = \beta\gamma$ satisfies (*).

Let J_i be the component of ∂F containing $x_i = \alpha(i)\,(i = 0, 1)$ and let K be the component of $\partial G'$ containing $y = f'(x_0) = f'(x_1)$.

Let $p : (\tilde{F}, \tilde{x}_0) \to (F, x_0)$ be the covering space with $p_*\pi_1(\tilde{F}, \tilde{x}_0) = \eta_0\pi_1(J_0, x_0)$ (where $\eta_0 : \pi_1(J_0, x_0) \to \pi_1(F, x_0)$ is inclusion induced). Let $\tilde{\alpha}$ be the lifting of α with $\tilde{\alpha}(0) = \tilde{x}_0$. Put $\tilde{x}_1 = \tilde{\alpha}(1)$ and let $\tilde{J}_i\,(i = 0, 1)$ be the component of $p^{-1}(J_i)$ containing $\tilde{x}_i$. Now $\tilde{J}_1 \neq \tilde{J}_0$; otherwise, since $\pi_1(\tilde{J}_0) \to \pi_1(\tilde{F})$ is an isomorphism, $\tilde{\alpha}$ is homotopic (end points fixed) to a path $\tilde{\alpha}_1$ in $\tilde{J}_0$. But then $f' \circ p \circ \tilde{\alpha}_1$ is a loop in K which lifts to a path under the covering $f'|J_0 : J_0 \to K$. Since $f' \circ p \circ \tilde{\alpha}_1$ is nullhomotopic in G', we contradict the fact that $\pi_1(K) \to \pi_1(G')$, is monic.

Consider the inclusion induced maps

$$\eta_i : \pi_1(J_i, x_i) \to \pi_1(F, x_i), \quad \tilde{\eta}_i : \pi_1(\tilde{J}_i, \tilde{x}_i) \to \pi_1(\tilde{F}, \tilde{x}_i)$$

and the change of base point maps

$$\psi_\alpha : \pi_1(F, x_1) \to \pi_1(F, x_0) \ ,$$

$$\psi_{\tilde{\alpha}} : \pi_1(\tilde{F}, \tilde{x}_1) \to \pi_1(\tilde{F}, \tilde{x}_0)$$

induced by α and $\tilde{\alpha}$. Note that

$$p_* \psi_\alpha \tilde{\eta}_1 \pi_1(\tilde{J}_1, \tilde{x}_1) = \psi_\alpha \eta_1 \pi_1(J_1, x_1) \cap \eta_0 \pi_1(J_0, x_0) \ .$$

Since $[f' \circ \alpha] = 1$, we have a commutative diagram

$$\begin{array}{ccccc}
\pi_1(J_1, x_1) & & \xrightarrow{\quad (f'|J_1)_* \quad} & & \\
\downarrow \psi_\alpha \circ \eta_1 & & & & \\
\pi_1(F, x_0) & \xrightarrow{f'_*} & \pi_1(G', y) & \longleftarrow & \pi_1(K, y) \ . \\
\uparrow \eta_0 & & & & \\
\pi_1(J_0, x_0) & & \xrightarrow{\quad (f'|J_0)_* \quad} & &
\end{array}$$

The maps $(f'|J_i)_*$, being induced by finite sheeted coverings, have images of finite index in $\pi_1(K, y)$. From this we conclude that the intersection $\psi_\alpha \eta_1 \pi_1(J_1, x_1) \cap \eta_0 \pi_1(J_0, x_0)$ has finite index in each term. Hence $\tilde{J}_1$ is compact, and a nontrivial power of any loop in $\tilde{J}_0$ is freely homotopic in $\tilde{F}$ to a loop in $\tilde{J}_1$. It must be that $\tilde{F}$ is an annulus. Thus F is an annulus on mobius band. Using the fact that $[f \circ \alpha] = 1$, one easily constructs a homotopy retracting f' into $\partial G'$ and conclusion (ii) follows.

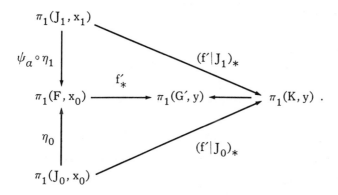

Now suppose that $f'|\partial F$ is an embedding. A simple homology argument shows that G' is compact. Let B be a properly embedded, nonseparating arc in G'. We may assume that $A = f'^{-1}(B)$ is a single arc in F and that $f'|A$ is an embedding. Letting G_1 be G' cut open along B, we may further assume that $f'|f'^{-1}(G_1)$ is boundary preserving and is an embedding when restricted to the boundary. Since $\beta_1(G_1) < \beta_1(G')$, we may use an inductive argument to show that $f'|f'^{-1}(G_1)$ is homotopic (rel ∂) to a homeomorphism. Thus f' is homotopic to a homeomorphism and conclusion (i) follows.

For the case $\partial G = \emptyset$, if $G = P^2$, then $F = P^2$ and conclusion (i) follows. Otherwise we can choose a 2-sided nonseparating simple closed B in G. By cutting open along G, we reduce to the bounded case. If conclusion (ii) holds for one of the resulting pieces, we can reduce the number of components of $f^{-1}(B)$. We cannot completely eliminate $f^{-1}(B)$; otherwise f_* would factor through the free group $\pi_1(G-B)$ and could not be monic. The conclusion (i) follows by fitting together the pieces.

Hierarchies

If M is a compact 3-manifold, a sequence

$$M = M_0 \supset M_1 \supset \cdots \supset M_n$$

of 3-submanifolds is called a *hierarchy* for M provided that M_{i+1} is obtained from M_i by cutting open along a properly embedded, 2-sided incompressible surface F_i and each component of M_n is a 3-cell. The integer n is called the *length* of the hierarchy.

The following lemma is useful in proving the existence of hierarchies (a different approach is given in [105]). More general versions of this lemma are given in [31] and [32] (see also [113] for some corrections).

13.2 LEMMA. *Let* M *be a compact, irreducible* 3-*manifold. There is an integer* n(M) *such that if* $\{S_1, \cdots, S_k\}$ *is any collection of* k *pairwise disjoint, closed,* 2-*sided, incompressible surfaces in* M *with* $k > n(M)$, *then some pair* S_i, S_j *must be parallel* (*cobound a product*) *in* M.

PROOF. This follows much the same line as the proof of Kneser's theorem (3.14) to which we refer for details.

We fix a triangulation T of M. Using the arguments of 3.14 and the incompressibility of the surfaces, any collection $\{S_1, \cdots, S_k\}$ can be put in "normal form" with respect to T in the sense that if τ is a 3-simplex of T, then each component, D, of $\tau \cap \cup S_i$ is a 2-cell and each component of $\tau - D$ contains a vertex of τ. The proof of this is accomplished by a sequence of isotopies and "disk replacements." By irreducibility, each "disk replacement" can also be achieved by an isotopy.

For τ, a 3-simplex of T a component, C, of $\tau - \cup S_i$ is *good* if $C \cap \partial\tau$ is an annulus containing no vertex of τ. Note that for a fixed τ, there can be at most six "bad" components.

Now we put $n(M) = \beta_1(M) + \beta_1(M; Z_2) + 6t$ where t is the number of 3-simplexes of T. If $k > n(M)$, then $M - \cup S_i$ has more than $\beta_1(M; Z_2) + 6t$ components. At most 6t of these can be "bad." As in 3.14, the closure of each good component is an I-bundle over some surface. At most, $\beta_1(M; Z_2)$ could be twisted I-bundles; so we have at least one product bundle.

13.3 THEOREM. *Let* M *be a compact, connected,* P^2*-irreducible 3-manifold with* $\partial M \neq \emptyset$. *Then* M *has a hierarchy*

$$M = M_0 \supset M_1 \supset \cdots \supset M_n$$

with each M_i *connected. If* M *is orientable, we may suppose that each* F_i *has nonempty boundary* ($M_{i+1} = M_i$ *cut open along* F_i).

PROOF. We construct a sequence $M = N_0 \supset N_1 \supset N_2 \supset \cdots$ of (not necessarily connected 3-manifolds) inductively by the following role. If some component of ∂N_i is compressible (and $\neq S^2$), let G_i be a 2-cell properly embedded in N_i with ∂G_i not contractible in ∂N_i. If every component of ∂N_i is incompressible, we let G_i be a properly embedded, 2-sided incompressible surface in N_i which doesn't separate the component of N_i which contains it (by 6.6 and 6.7 G_i exists). We let N_{i+1} be N_i cut open along G_i and denote the parallel copies of G_i in ∂N_{i+1} by G'_i and G''_i. We may further assume that $G_i \cap (G'_j \cup G''_j) = \emptyset$ if $j < i$ and G_j is a 2-cell.

We claim that this process must stop at some stage N_m -- in which case (by P^2-irreducibility) every component of N_m must be a 3-cell. For suppose the process could be continued indefinitely. Noting that whenever G_i is a 2-cell, ∂N_{i+1} "simpler" than ∂N_i, we must have arbitrarily long sequences $N_{i_1} \supset N_{i_2} \supset \cdots$ in which each component of ∂N_{i_j} is incompressible in N_{i_j}. Since only finitely many of the G_i can be closed, we may assume each G_{i_j} has nonempty boundary. Let S_j be the component of ∂N_{i_j} which meets G_{i_j}. By pushing S_j slightly into Int N_{i_j}, we may assume that $S_1, S_2 \cdots$ are pairwise disjoint. Let $n(M)$ be the integer given by 13.2. It follows that for some $j < k \leq n(M) + 1$, S_j is parallel to S_k. This is impossible since S_k lies in N_{i_j} cut open along G_{i_j} and $S_j \cap G_{i_j} \neq \emptyset$.

This justifies our claim. So the process indeed stops, say, at N_m -- each component of N_m is a 3-cell. Now $N_0 = M$ is connected and N_{i+1} has more components than N_i precisely when G_i is a 2-cell which separates the component of N_i in which it lies. The proof is completed by noting that we could omit these separating 2-cells. For let G_j be the last one of them. Then G'_j and G''_j meet no G_i with $i > j$; hence G'_j and G''_j are contained in ∂N_n. They in fact lie in different components of N_m. Thus the effect of omitting the jth cut is simply to reduce the number of components of N_m. By repeating this argument, we may omit all the separating 2-cells and the proof is complete.

13.4 THEOREM. *Let* $M \in \mathfrak{W}$. *Then the universal cover of* M *embeds in* R^3. *The universal cover of* Int M *is an open* 3-*cell.*

PROOF. By 13.3, there is a hierarchy $M = M_0 \supset M_1 \supset \cdots \supset M_n$ for M (it $\partial M = \emptyset$, we first cut along an incompressible surface and then apply 13.3). Let $p : \tilde{M} \to M$ be the universal cover. Since $\pi_1(M_1) \to \pi_1(M)$ and $\pi_1(F_0) \to \pi_1(M)$ are monic, for each component $\tilde{M}_1$ of $p^{-1}(M_1)$, $p|\tilde{M}_1 : \tilde{M}_1 \to M_1$ is the universal cover of M_1. Also, each component of $p^{-1}(\text{Int } F_0)$ is an open 2-cell.

By inductively arguing the conclusion for M_1, we obtain the conclusion for M by piecing together the components of $p^{-1}(M_1)$. Some care must be taken since $p^{-1}(M_1)$ will have infinitely many components. We leave the details to the reader.

13.5 COROLLARY. *Let* $M \in \mathfrak{W}$, *then any covering of* M *is* p^2*-irreducible.*

PROOF. By the sphere and projective plane theorems, a cover of M could fail to be p^2-irreducible only if it contained a fake 3-cell. However, a homotopy 3-cell would lift to the universal cover and the conclusion follows from 13.4.

Classification Theorems

13.6. THEOREM. *Let* $M, N \in \mathfrak{W}$ *and suppose* $f : (M, \partial M) \to (N, \partial N)$ *is a map such that* $f_* : \pi_1(M) \to \pi_1(N)$ *is monic and such that for each component* B *of* ∂M $(f|B)_* : \pi_1(B) \to \pi_1(C)$ *is monic, where* C *is the component of* ∂N *containing* $f(B)$. *Then there is a homotopy* $f_t : (M, \partial M) \to (N, \partial N)$ *such that* $f_0 = f$ *and either*

(i) $f_1 : M \to N$ *is a covering map,*

(ii) M *is an I-bundle over a closed surface, and* $f_1(M) \subset \partial N$, *or*

(iii) N *(hence also* M*) is a solid torus or a solid Klein bottle and* $f_1 : M \to N$ *is a branched covering with branch set a circle.*

If $f|B : B \to C$ *is already a covering map, we may assume* $f_t|B = f|B$ *for all* t.

PROOF. Note the condition that $(f|B)_* : \pi_1(B) \to \pi_1(C)$ be monic is automatically satisfied if the components of ∂M are incompressible in M (or if $\partial M = \emptyset$).

By 13.1 we may assume, after changing f by a homotopy, that for each component, B, of ∂M, $f|B : B \to f(B)$ is a covering map. If this is already the case for some B, there is no need now nor in any future step to change $f|B$.

We construct a commutative diagram

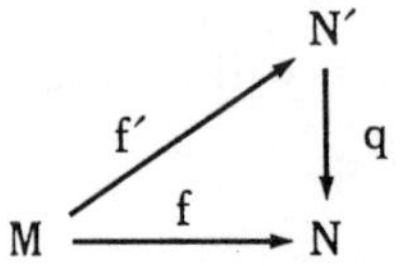

where $q : N' \to N$ is a covering map and $f'_* : \pi_1(M) \to \pi_1(N')$ is an isomorphism. Clearly $f'|B : B \to f'(B)$ is a covering map for each component B of ∂M.

First consider the case in which $\partial M \neq \emptyset$ and $f'|\partial M$ is *not* an embedding. Let x_0, x_1 be a pair of distinct points of ∂M with $f'(x_0) = f'(x_1)$. Since f'_* is epic, there is a path $\alpha : (I, \partial I) \to (M, \partial M)$ satisfying

(∗) $\alpha(i) = x_i$ $(i = 0, 1)$ (hence $f'(\alpha(0)) = f'(\alpha(1))$) and the loop $f' \circ \alpha$ is nullhomotopic in N'.

Let B_i be the component of ∂M containing x_i and C be the component of $\partial N'$ containing $y = f(x_0) = f(x_1)$. Let $p : (\tilde{M}, \tilde{x}_0) \to (M, x_0)$ be the covering space with $p_* \pi_1(\tilde{M}, \tilde{x}_0) = \eta_0 \pi_1(B_0, x_0)$. Let $\tilde{\alpha}$ be a lifting of α with $\tilde{\alpha}(0) = \tilde{x}_0$, let $\tilde{x}_1 = \tilde{\alpha}(1)$ and let $\tilde{B}_i$ be the component of $p^{-1}(B_i)$ containing $\tilde{x}_i$.

As in the proof of 13.1, we have a commutative diagram

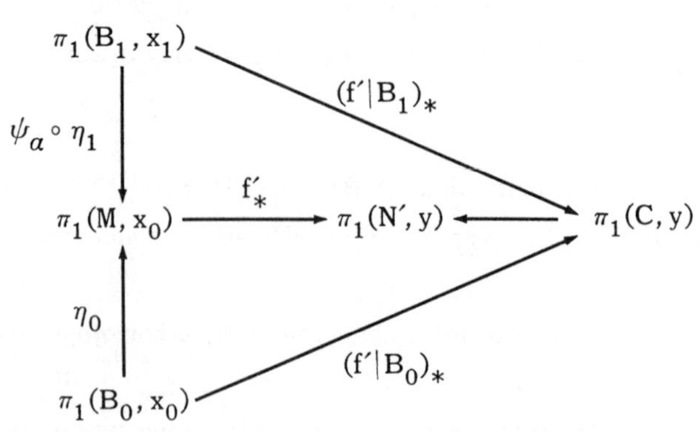

where the η's are inclusion induced and ψ_α is the change of base point map induced by α. Commutativity follows from the fact that $[f' \circ \alpha] = 1$.

Since $(f'|B_i)$ is a finite sheeted covering, we again conclude that $\psi_\alpha \eta_1 \pi_1(B_1, x_1) \cap \eta_0 \pi_1(B_0, x_0)$ has finite index in each term. Since this intersection is $p_* \psi_\alpha \tilde{\eta}_1 \pi_1(\tilde{B}_1, \tilde{x}_1)$, we conclude that $\tilde{B}_1$ is compact and that a nonzero power of each loop in $\tilde{B}_0$ is freely homotopic in $\tilde{M}$ to a loop in B_1. Note also that $p|\tilde{B}_0 : \tilde{B}_0 \to B_0$ is a homeomorphism.

Suppose that α also satisfies

(**) α is not homotopic (rel ∂) to a path in ∂M (this is certainly the case if $B_0 \neq B_1$).

With this assumption, we must have $\tilde{B}_0 \neq \tilde{B}_1$; otherwise, since $\pi_1(\tilde{B}_0) \to \pi_1(\tilde{M})$ is epic, $\tilde{\alpha}$ would homotope into $\tilde{B}_0$ and projecting this homotopy would contradict (**). In addition, we can conclude that $\tilde{B}_0$ is incompressible in $\tilde{M}$; if not, we could write $\pi_1(\tilde{M})$ as a free product with the image of $\pi_1(\tilde{B}_1)$ conjugate to a subgroup of one factor. This is not possible since $\pi_1(\tilde{B}_1)$ maps to a subgroup of finite index in $\pi_1(\tilde{M})$. Thus $\tilde{\eta}_0 : \pi_1(\tilde{B}_0) \to \pi_1(\tilde{M})$ is an isomorphism, and $\tilde{\eta}_0$ is a homotopy equivalence. From

$$0 \to H_3(\tilde{M}, \tilde{B}_0 \cup \tilde{B}_1; Z_2) \to \underset{\substack{\wr\wr \\ Z_2 + Z_2}}{H_2(\tilde{B}_0 \cup \tilde{B}_1; Z_2)} \to \underset{\substack{\wr\wr \\ Z_2}}{H_2(\tilde{M}; Z_2)}$$

we conclude that $\tilde{M}$ is compact; hence $\eta_0 \pi_1(B_0)$ has finite index in $\pi_1(M)$. By 10.6, M is an I-bundle over a closed surface. This gives conclusion (ii) of the theorem. To obtain the homotopy retracting M into ∂N, consider the covering $q' : N'' \to N'$ corresponding to $f'_* \pi_1(B_0)$. An appropriate lifting f'' of f' takes B_0 and B_1 into a component of $q^{-1}(C)$ (the same component since $[f' \circ \alpha] = 1$) which is a deformation retract of N' by a homotopy $\rho_t : N'' \to N''$, then $f_t = q \circ q' \circ \rho_t \circ f''$ is the desired homotopy.

Now we suppose there is no path α satisfying both (*) and (**). If $f'|\partial M$ is not an embedding, we have a path α satisfying (*), but α is homotopic (rel ∂) to a path $\alpha_1 : I \to \partial M$ (hence $B_0 = B_1$). The loop $f' \circ \alpha_1$ is not contractible in C; since it does not lift to a loop under the covering map $f'|B_0 : B_0 \to C$. However, $f' \circ \alpha_1$ is null homotopic in N'. Thus C is compressible in N'. We wish to show that C is a torus or Klein bottle and deduce conclusion (iii) of the theorem. For this it is convenient to reduce to the orientable case. So consider any commutative diagram

$$\begin{array}{ccc} \overline{M} & \xrightarrow{\ \overline{f}\ } & \overline{N} \\ {\scriptstyle \overline{p}}\downarrow & & \downarrow{\scriptstyle \overline{q}} \\ M & \xrightarrow{\ f'\ } & N' \end{array}$$

where $\overline{p}$ and $\overline{q}$ are covering maps, with $\overline{p}$ finite sheeted, $\overline{M}$ and $\overline{N}$ orientable, and $\overline{f}_* : \pi_1(\overline{M}) \to \pi_1(\overline{N})$ is an isomorphism.

If f maps two distinct components of $\partial\overline{M}$ to the same component of $\partial\overline{N}$, then there is a path β joining these components such that $\overline{f} \circ \beta$ is a loop. Since $\overline{f}_*$ is epic, we may assume $[\overline{f} \circ \beta] = 1$. But then $\overline{p} \circ \beta$ satisfies (*) and (**) and we are back to the previous case. So we may suppose that $\overline{f}$ takes distinct components of $\partial\overline{M}$ to distinct components of $\partial\overline{N}$. If $\overline{B}_0$ is a component of $\overline{p}^{-1}(B_0)$, then $\overline{f}|\overline{B}_0$ is not one to one. This is because a lifting of α which begins in $\overline{B}_0$ must also end in $\overline{B}_0$ whereas $f' \circ \alpha$ lifts to a loop.

Consider the following diagram (integer coefficients)

$$\begin{array}{ccc} 0 & & 0 \\ \downarrow & & \downarrow \\ H_3(\overline{M}, \partial\overline{M}) & \xrightarrow{\ \ \overline{f}_*\ \ } & H_3(\overline{N}, \partial\overline{N}) \\ {\scriptstyle \partial_*}\downarrow & & \downarrow{\scriptstyle \partial_*} \\ H_2(\partial\overline{M}) & \xrightarrow{\ (\overline{f}|\partial\overline{M})_*\ } & H_2(\partial\overline{N}) \end{array} .$$

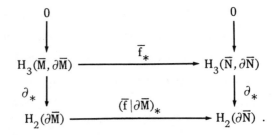

By the above remarks $(\overline{f}|\partial\overline{M})_* \circ \partial_* \neq 0$. Thus $\overline{f}_* \neq 0$, and it follows that $\overline{N}$ is compact and that every component of $\partial\overline{N}$ is the image, under $\overline{f}$, of some component of $\partial\overline{M}$.

Now $\overline{f}$ is a homotopy equivalence, so

$$\tfrac{1}{2}\chi(\partial\overline{M}) = \chi(\overline{M}) = \chi(\overline{N}) = \tfrac{1}{2}\chi(\partial\overline{N}) .$$

If $\overline{M}$ has components $\overline{B}_0,\cdots,\overline{B}_k$, then we have

$$\chi(\partial\overline{M}) = \sum \chi(\overline{B}_i) = \sum n_i\,\chi(\overline{f}(\overline{B}_i)) = \sum \chi(\overline{f}(\overline{B}_i)) = \chi(\partial\overline{N}) ,$$

where

$$\overline{f}\,|\,\overline{B}_i : \overline{B}_i \to \overline{f}(\overline{B}_i)$$

is n_i-sheeted. Thus $n_i = 1$ unless $\chi(\overline{f}(\overline{B}_i)) = 0$. Since $n_0 > 1$, $\chi(\overline{f}(\overline{B}_0)) = 0$. So C is a compressible torus or Klein bottle. Thus N' (hence also M, N) is a solid torus or solid Klein bottle and we obtain conclusion (iii). This completes the proof in case $f'|\partial M$ is not an embedding.

Now assume that $f'|\partial M$ is an embedding (automatic if $\partial M = \emptyset$). Let $N' = N'_0 \supset N'_1 \supset \cdots \supset N'_n$ be a hierarchy for N'. We prove, by induction on n, that f' is homotopic ($f'|\partial M$ unaltered) to a covering map (a homeomorphism if $\partial M \neq \emptyset$). Composing with $q : N' \to N$ gives conclusion (i).

If $n = 1$, then N' (also M) is a solid torus or a solid Klein bottle. The homeomorphism $f'|\partial M : \partial M \to \partial N'$ extends to a homeomorphism of M to N' which is homotopic to f'.

Suppose that $n > 1$. Let G be the incompressible surface in N' along which we cut to obtain N'_1. We may change f' by a homotopy (without changing $f'|\partial M$) so that each component of $f'^{-1}(G)$ is properly embedded, 2-sided and incompressible. If F is a component of $f'^{-1}(G)$, then $(f'|F)_* : \pi_1(F) \to \pi_1(G)$ is monic. Thus if $\partial G \neq \emptyset$, no component of $f'^{-1}(G)$ is closed; so $f'|f^{-1}(G) : f^{-1}(G) \to G$ is boundary preserving and thus is an embedding on $\partial f^{-1}(G)$. If G is a 2-cell, then so is $f^{-1}(G)$

and we may homotope $f'|f^{-1}(G)$ to an embedding. In any other case 13.1 applies; so we may homotope $f'|f'^{-1}(G)$ to a covering map. The exceptional cases of 13.1 cannot occur since $f'|\partial f'^{-1}(G)$ is an embedding (a Möbius band doesn't retract to its boundary). For the same reason, we must have (the modified) $f'|f'^{-1}(G)$ an embedding if $\partial G \neq \emptyset$. Let Q be a component of N_1. By the above, we may assume that if P is a component of $f'^{-1}(Q)$, then $f' : P \to Q$ is boundary preserving and f' is a covering map on each component of ∂P. By incompressibility, $(f'|P)_* : \pi_1(P) \to \pi_1(Q)$ is monic.

If $f'|\partial P$ is an embedding (which must occur if $\partial G \neq \emptyset$), then we apply induction to homotope $f'|P$ to a homeomorphism.

If $f'|\partial P$ is not an embedding (so G is closed), then our initial considerations apply to show that P is an I-bundle over a closed surface and that $f'|P$ homotops into G. We note that conclusion (iii) does not occur here, since P contains a closed incompressible surface (parallel to a component of $f'^{-1}(G)$) in its boundary. Since $P \subset \text{Int } M$, we may eliminate (one or two) components of $f'^{-1}(G)$. Continuing in this manner, we must arrive at a point where f' is an embedding on the boundary of each component of $f'^{-1}(N_1)$. Applying induction and piecing together the results, we see that f' is homotopic to a covering map. If $\partial M \neq \emptyset$, then since $f'|\partial M$ is an embedding, f' is homotopic to a homeomorphism. This completes the proof of 13.6.

13.7 COROLLARY. *Let* $M, N \in \mathfrak{W}$ *and suppose* $f : (M, \partial M) \to (N, \partial N)$ *is a map such that* $f_* : \pi_1(M) \to \pi_1(N)$ *is an isomorphism and such that for each component* B *of* ∂M $(f|B)_* : \pi_1(B) \to \pi_1(C)$ *is monic* (C *the component of* ∂N *containing* f(B)). *Then either* f *is homotopic to a homeomorphism of* M *to* N *or* M *is a twisted* I-*bundle over a closed surface and* N *is a product* I-*bundle over a homeomorphic surface.*

PROOF. Note that we do not require the homotopy to preserve boundary.

If $M = B^3$, the conclusion is trivial. Otherwise we can apply 13.6. Conclusion (i) gives the desired result. In conclusion (ii), N is also an

I-bundle. In fact, N is a product bundle; otherwise $f_*\pi_1(M)$ is contained in $\pi_1(\partial N)$ which has index two in $\pi_1(N)$. If M is also a product bundle, there is clearly a homeomorphism of M to N homotopic to f. In conclusion (iii) we again can construct a homeomorphism of M to N homotopic to f.

Peripheral Systems

The hypothesis of 13.6 and 13.7 can be reduced to purely group theoretic statements. This is because we are dealing with aspherical manifolds; so any homomorphism of fundamental groups is induced by a continuous map. However, we require boundary preserving maps; so some additional care is needed – an arbitrary homomorphism of fundamental groups is not induced by a boundary preserving map (cf. examples 13.11 at the end of this chapter).

For this purpose let M be a 3-manifold with boundary components $B_1, \cdots, B_m$. The *peripheral group system* of M is the collection

$$\{\pi_1(M); \eta_i : \pi_1(B_i) \to \pi_1(M), i = 1, \cdots, m\}$$

consisting of the fundamental group of M together with the inclusion induced maps, η_i, of the fundamental groups of the boundary components. Note that the η_i involve base point considerations in their definition and are therefore determined up to inner automorphism of $\pi_1(M)$. We will use the notation, without additional adornments, with the understanding that this ambiguity occurs.

Let N be another 3-manifold with boundary components $C_1, \cdots, C_n$ and inclusion induced maps $\theta_j : \pi_1(C_j) \to \pi_1(N)$. We say that a homomorphism $\phi : \pi_1(M) \to \pi_1(N)$ *preserves the peripheral structure* provided that for $i = 1, \cdots, n$, there is an integer $j(i)$, a homomorphism $\psi_i : \pi_1(B_i) \to \pi_1(C_{j(i)})$, and an inner automorphism $\alpha_i : \pi_1(N) \to \pi_1(N)$ such that

$$\phi \circ \eta_i = \alpha_i \circ \theta_i \circ \psi_i : \pi_1(B_i) \to \pi_1(N) .$$

The collection $\{\phi; \psi_1, \cdots, \psi_m\}$ is called a *homomorphism of peripheral group systems*. Such a collection is called a *monomorphism of peripheral group systems* if ϕ *and* each ψ_i is a monomorphism.

If each η_i and each θ_j is monic, the definition becomes simpler: *If* M *and* N *have incompressible boundary components, then a homomorphism* $\phi : \pi_1(M) \to \pi_1(N)$ *preserves the peripheral structure if for each* i, $\phi(\eta_i(\pi_i(B_i))$ *is conjugate to a subgroup of some* $\theta_j(\pi_1(C_j))$. However, in general we must have the homomorphisms ψ_i as defined.

Clearly a boundary preserving map $f : (M, \partial M) \to (N, \partial N)$ induces the homomorphism of peripheral group systems $\{f_*; (f|B_i)_*\}$. In case $\pi_2(N) = 0$, the converse is also true.

13.8 LEMMA. *Let* M, N *be* 3*-manifolds with* $\pi_2(N) = 0$, *then any homomorphism of the peripheral group system of* M *to that of* N *is induced by a boundary preserving map*

$$f : (M, \partial M) \to (N, \partial N) \ .$$

PROOF. The ψ_i's are used to define f on ∂M. The inner automorphism α_i indicates how to extend over a tree, T, in M which joins the boundary components of M to a common base point. One enlarges T to a maximal tree, Γ, relative to ∂M (one includes no more vertices of ∂M) and extends over Γ by retracting to T. The extension over the remaining 1-simplexes is determined by ϕ. The rest of the proof is standard – details are left to the reader.

13.9 THEOREM. *Let* $M.N \in \mathfrak{W}$. *If there is a monomorphism of the peripheral group system of* M *to that of* N, *then there is a boundary preserving map* $f : (M, \partial M) \to (N, \partial N)$ *inducing this monomorphism. Thus the conclusion of* 13.6 (*and, if* $f_* : \pi_1(M) \to \pi_1(N)$ *is epic,* 13.7) *follows.*

Remarks and Examples

There are two directions in which the above results may be generalized.

First the assumption, in 13.6, that $(f|B)_*: \pi_1(B) \to \pi_1(C)$ be monic for each component B of ∂M is somewhat awkward. It is natural to ask what happens under the assumption:

For $M, N \in \mathfrak{W}$, there is a map $f: (M, \partial M) \to (N, \partial N)$ with $f_*: \pi_1(M) \to \pi_1(N)$ monic.

Evans [22] proves, in the orientable case, an analogue of Corollary 13.7 with this weaker assumption. His proof is surprisingly involved. It is plausible that, properly stated, 13.6 remains valid under the weaker assumption:

13.10 CONJECTURE. *With the above assumption, the conclusion of* 13.6 *remains valid if we add the possibility*

(iv) M *is a cube with handles and* $f_1(M) \subset \partial N$.

We outline a "proof." By 13.6 we may assume that for some component B of ∂M, $(f|B)_*: \pi_1(B) \to \pi_1(C)$ is not monic. Suppose that some nontrivial element of $\ker(f|B)_*$ is represented by a simple closed curve J. Then J is contractible in M. By Dehn's lemma, J bounds a 2-cell D in M. Let M_1 be M cut open along D. Since $f|\partial D$ is contractible in C and $\pi_2(N) = 0$, $f|M_1$ is homotopic to a boundary preserving map $f': (M_1, \partial M_1) \to (N, \partial N)$. We can apply an induction to (the components of) M_1. If conclusion (i) or (iii) holds for a component M_1' of M_1, then $f'_*(\pi_1(M_1'))$ has finite index in $\pi_1(N)$. This is impossible since $\pi_1(M_1')$ has infinite index in $\pi_1(M)$. In the other cases, f homotops (through boundary preserving maps) to a map $f_1: M \to C$. If, in fact, conclusion (iv) holds for each component of M_1, then conclusion (iv) holds for M. If conclusion (ii) holds for a component M_1' of M, then C must be incompressible in N and $(f_1|M_1')_*(\pi_1(M_1'))$ has finite index in $\pi_1(C)$. This is again impossible.

The problem with this "proof" is finding the *simple* closed curve J. It is not true, in general, that for a map $f: B \to C$ of surfaces that $\ker f_*$, if nontrivial, "contains" a simple closed curve. However, there is more information available here and it may be possible to make this "proof" valid.

A second direction of generalization is in weakening the assumption of irreducibility. Suppose, for convenience, that M and N are closed and contain no fake 3-cells. Then by Kneser's Theorem (7.1), an isomorphism between $\pi_1(M)$ and $\pi_1(N)$ induces isomorphisms between the fundamental groups of the prime factors of M and N respectively. The nonirreducible prime factors are 2-sphere bundles over S^1 and are characterized by having infinite cyclic fundamental groups. If we have no 2-sided P^2's (say M and N are orientable), then by 13.7, the sufficiently large prime factors are pairwise homeomorphic. Thus the obstructions to finding a homeomorphism between M and N are

(1) the existence of nonsufficiently large prime factors, and

(2) the fact that $M_1 \# M_2 \neq M_1 \# (-M_2)$ can occur (3.22).

We conclude by giving some examples showing that isomorphism of fundamental groups alone is not sufficient to imply homeomorphism. Examples occur among knot spaces of composite knots (cf. [26]). The following provides examples of manifolds with the same fundamental group as a torus knot space, but which do not embed in S^3.

13.11 EXERCISE. *Let* $T_i (i = 1, 2)$ *be a solid torus. Choose a standard longitude meridian pair* (λ_i, μ_i) *in* ∂T_i. *For* $p_i s_i - q_i r_i = +1$, *let* $a_i = \lambda_i^{p_i} \mu_i^{q_i}$, $\beta_i = \lambda_i^{r_i} \mu_i^{s_i}$. *Identify an annular neighborhood of* a_1 *with an annular neighborhood of* a_2 *by a homeomorphism which takes* a_1 *to* a_2 *in an orientation preserving manner and reverses orientation on a transverse arc to obtain* M. *Assuming that* $|p_i| > 1$, *show that* M *embeds in* S^3 *if and only if* $q_1 \equiv \varepsilon p_2 \bmod p_1$ *and* $q_2 \equiv \varepsilon p_1 \bmod p_2$ *for* $\varepsilon = \pm 1$ *(same sign in both congruences).*

HINT. Observe that $\pi_1(M) = \langle \lambda_1, \lambda_2 : \lambda_1^{p_1} = \lambda_2^{p_2} \rangle$ and that $a = \lambda_1^{p_1} = \lambda_2^{p_2}$ and $\beta = \beta_1 \beta_2 = \lambda_1^{r_1} \lambda_2^{r_2}$ serve as standard generators for $\pi_1(\partial M)$. M embeds in a homotopy 3-sphere if and only if it is possible to obtain the trivial group from $\pi_1(M)$ by adding a relation

$$(\lambda_1^{p_1})^n (\lambda_1^{r_1} \lambda_2^{r_2})^k = 1$$

where $(n, k) = 1$.

Let $G = <\lambda_1, \lambda_2 : \lambda_1^{p_1} = \lambda_2^{p_2}, \lambda_1^{np_1}(\lambda_1^{r_1} \lambda_2^{r_2})^k = 1>$. Note that $G/G' = 1$ if and only if $k(p_1 r_2 + p_2 r_1) - np_1 p_2 = \pm 1$.

Let

$$H = < R_1, R_2, R_3 : R_1^2 = R_2^2 = R_3^2 = (R_1 R_2)^{p_1} = (R_2 R_3)^{p_2} = (R_1 R_3)^k = 1 > .$$

If, $|p_1|, |p_2|, |k| \geq 2$, then H can be interpreted as a group of rigid motions of the 2-sphere, the plane, or the hyperbolic plane according as $1/|p_1| + 1/|p_2| + 1/|k|$ is greater than, equal to, or less than one; the R_i's correspond to reflections about the sides of a triangle with angles $\pi/|p_1|$, $\pi/|p_2|$, and $\pi/|k|$.

Show that $\lambda_1 \to (R_1 R_2)^{-q_1}$, $\lambda_2 \to (R_2 R_3)^{-q_2}$ induces a nontrivial representation of G unless $k = \pm 1$. Use the homology conditions to further deduce that if $G = 1$, then $q_i \equiv \varepsilon p_j \bmod p_i$. If these conditions hold, one can explicitly construct the embedding of M into S^3.

CHAPTER 14
SOME APPROACHES TO THE POINCARÉ CONJECTURE

In some ways the unsettled nature of the Poincaré conjecture has not been an impediment to the development of the theory. This is evidenced by the number of theorems which characterize the Poincaré associate of a 3-manifold in terms of its fundamental group. In other ways this is definitely not the case. For example, if $\pi_1(M)$ is finite, then, even if M is irreducible, the universal cover of M might be a homotopy 3-sphere – there is no way of bypassing the Poincaré conjecture in this case.

In any event, the Poincaré conjecture remains a fundamental and intriguing problem in the subject. We wish to give some insight to the nature of this problem by surveying some of the attempts to prove (and to disprove) it, and by noting some equivalent formulations of the problem in purely algebraic terms.

If M is any closed orientable 3-manifold, then

1. M can be obtained from S^3 by surgery on some link [59], [107], and

2. M is a branched cover of S^3 with branch set some knot in S^3 [1], [44], [72], and

3. M has a Heegaard splitting.

Each of these constructions provides an approach to the Poincaré conjecture.

There are numerous classes of knots, K, in S^3 which are known to have the property that no surgery on K can produce a counterexample to the Poincaré conjecture. These include torus knots, twist knots, almost all cable knots, double knots, composite knots, and many pretzel knots.

See [8], [94] for a summary of results in this direction. However, there is little in the way of general results using a surgery approach to the Poincaré conjecture.

Fox [27] suggested that the branched coverings of S^3 provide a good source of possible counterexamples to the Poincaré conjecture. This approach is receiving renewed attention for the following reasons. Any attempt to establish a counterexample must involve a property or invariant which can distinguish among manifolds of the same homotopy type. Such invariants are rare. One possibility is the μ-invariant, $\mu(M)$, which can be defined (cf. [46]) for any Z_2-homology 3-sphere M as the reduction modulo 16 of the signature of the intersection pairing $H_2(W)/_T \times H_2(W)/_T \to Z$ where W is any framed 4-manifold with $\partial W = M$ and T is the torsion subgroup of $H_2(W)$. Cappell and Shaneson [14] have developed a formula for computing $\mu(M)$ when M is a dihedral branched covering of S^3 branched over a knot. This formula is intrinsic in the sense that it only involves the description of M as a branched cover of S^3 and does not require construction of a 4-manifold bounded by M. Hilden [44] and, independently, Montesinos [72] have shown that every closed, orientable 3-manifold is, in fact, a 3-fold dihedral branched cover of S^3 branched over a knot. It should be noted that although a homotopy 3-sphere M with $\mu(M) \neq 0$ would be a counterexample to the Poincaré conjecture, that $\mu(M) = 0$ if $M \times S^1 \times S^1$ is p.l. homeomorphic to $S^2 \times S^1 \times S^1$ [14]; so the converse may not be true.

There has been considerable effort in using geometric techniques to "simplify" a Heegaard diagram for a homotopy 3-sphere; see [33] and [34].

We will concentrate on some reductions of the Poincaré conjecture to algebraic problems. In the final section we offer another method for seeking counterexamples.

Contractible Open 3-Manifolds

We first present examples to show that the obvious extension of the Poincaré conjecture to noncompact 3-manifolds is false.

14.1 EXAMPLES. *Consider the collection of open 3-manifolds,* M, *which can be written in the form* $M = \bigcup_{i=0}^{\infty} T_i$ *where each* T_i *is a solid torus and for each* $i \geq 0$

(i) $T_i \subset \text{Int } T_{i+1}$

(ii) $\pi_1(\partial T_{i+1}) \to \pi_1(T_{i+1} - T_i)$ *is monic, and*

(iii) T_i *is contractible in* T_{i+1}.

For one such example (J. H. C. Whitehead [108]), let T_0, T_1 be embedded in R^3 as shown below.

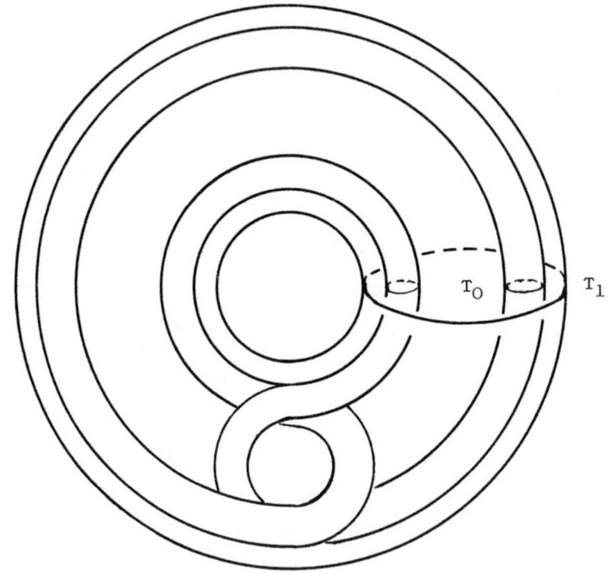

Both T_0 and T_1 are unknotted in R^3, so there is a homeomorphism $h: R^3 \to R^3$ such that $h(T_0) = T_1$. Put $T_i = h^i(T_0)$ and let $M = \cup T_i$. Note that, by construction, $M \subset R^3$. By varying this technique, an uncountable collection of pairwise nonhomeomorphic examples in R^3 is constructed in [68]. In [56], another uncountable collection of examples, none of which embeds in R^3, is constructed.

The interest in these examples is

14.2 THEOREM. *If* M *satisfies the conditions of* 14.1, *then* M *is a contractible open* 3*-manifold which is not homeomorphic to* R^3.

PROOF. Any compact subset of M lies in some T_i and is thus contractible in T_{i+1}. Thus $\pi_q(M) = 0$ for $q \geq 0$, and M is contractible.

To show that M is not homeomorphic to R^3, note that condition (ii) of 14.1, together with the loop theorem, shows that $\pi_1(\partial T_i) \to \pi_1(T_{i+1} - \text{Int}\, T_i)$ is also monic. Thus for $i < j$, $\pi_1(\partial T_j) \to \pi_1(T_j - T_i)$ is monic. If M is homeomorphic to R^3, then T_0 lies in a 3-cell C in M. But $C \subset \text{Int}\, T_j$ for some j. This is a contradiction since $\pi_1(\partial T_j) \to \pi_1(T_j - C)$ is not monic.

We should note that the above phenomenon is definitely related to noncompactness. The one point compactification of these examples is never a manifold; so we have not in any way jeopardized the status of the Poincaré conjecture.

A Characterization of S^3

The property used to distinguish the above examples from R^3 can be turned around to give a (non-homotopy-theoretic) characterization of S^3 [7]:

14.3 THEOREM. *Let* M *be a closed* 3*-manifold. Suppose that each simple closed curve in* M *lies in a* 3*-cell in* M. *Then* M *is homeomorphic to* S^3.

PROOF. The hypothesis clearly imply that $\pi_1(M) = 1$. In particular, M is orientable.

Let (V_1, V_2) be a Heegaard splitting (2.5) of M whose genus, g, is minimal. If $g = 0$, then $M = S^3$.

In any event, we may consider V_1 as an I-bundle over a surface, F, whose boundary consists of a single simple closed curve, J, and with $\chi(F) = 1 - g$. If g is even, F is orientable and V_1 is a product bundle; otherwise F is nonorientable and V_1 is a twisted I-bundle. We consider F as embedded in V_1 as a cross section to this bundle.

We claim that Int F is incompressible in $M-J$. If not, there is a 2-cell $D \subset M-J$ with $D \cap F = \partial D$ not contractible in F. We may assume that $D \cap V_1$ is an annulus. If ∂D does not separate F, then we obtain a cube with (g–1)-handles, V'_2, by cutting V_2 along $D \cap V_2$. Clearly $V'_1 = M - \text{Int } V'_2$ is a regular neighborhood of $F \cup D$. Now F collapses to a wedge of g circles. We may include the nonseparating curve ∂D as one of these circles. Thus $F \cup D$ collapses to a wedge of (g–1) circles and V'_1 is also a cube with (g–1)-handles. This gives the contradiction that (V'_1, V'_2) is a splitting of M of genus g–1. If ∂D separates F, then D together with some subsurface, F_1, of F bounds a cube with handles, W, in V_2. Let E be a properly embedded, nonseparating 2-cell in W. We may assume $\partial E \subset F_1$. Then ∂E does not separate F, so we get a contradiction as above. This justifies our claim that Int F is incompressible in $M-J$.

By hypothesis, there is a 3-cell $C \subset M$ with $J \subset \text{Int } C$. Using the irreducibility of $M-F$ and the incompressibility of F, we may assume that $F \subset C$. Thus $M - \text{Int } C$ is also a 3-cell and the proof is complete.

Splitting Homomorphisms

There are several ways of reformulating the Poincaré conjecture in purely algebraic terms. A number of these involve the idea of a *splitting homomorphism* [98]. Let (V_1, V_2) be a Heegaard splitting of a closed, orientable 3-manifold M, and let $T = \partial V_1 = \partial V_2$. We have a commutative diagram

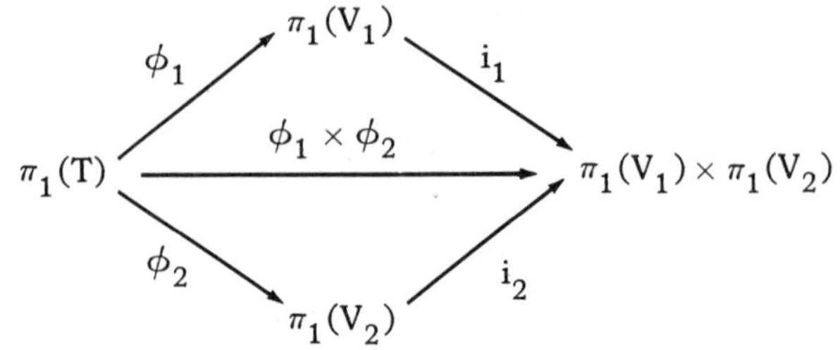

where ϕ_i is induced by inclusion. The map $\phi_1 \times \phi_2 : \pi_1(T) \to \pi_1(V_1) \times \pi_1(V_2)$ is called the *splitting homomorphism* associated with the given Heegaard splitting.

14.4 LEMMA. *With the above notation,* $\pi_1(M) = 1$ *if and only if* $\phi_1 \times \phi_2$ *is an epimorphism.*

PROOF. By Van Kampen's theorem, $j_* : \pi_1(T) \to \pi_1(M)$ is epic and $\ker j_* = \ker \phi_1 \cdot \ker \phi_2$. Since each of ϕ_1, ϕ_2 is epic, it follows that $\phi_1 \times \phi_2$ is epic if and only if $\ker \phi_1 \cdot \ker \phi_2 = \pi_1(T)$, and the proof is complete.

We note that $\pi_1(V_i)$ is free of rank g (= genus T). For each g we fix a free group F_g of rank g and view a splitting homomorphism as a map $\phi_1 \times \phi_2 : \pi_1(T) \to F_g \times F_g$. We must regard two such maps as equivalent if there is a commutative diagram

$$\begin{array}{ccc}
\pi_1(T) & \xrightarrow{\phi_1 \times \phi_2} & F_g \times F_g \\
\downarrow{\scriptstyle \alpha} & & \downarrow{\scriptstyle \beta_1 \times \beta_2} \\
\pi_1(T') & \xrightarrow{\phi'_1 \times \phi'_2} & F_g \times F_g
\end{array}$$

with α, β_1, and β_2 isomorphisms. The following result is due to Jaco [48].

14.5 LEMMA. *For* T *an orientable surface of genus* g, *every map* $\phi_1 \times \phi_2 : \pi_1(T) \to F_g \times F_g$ *with* ϕ_1 *and* ϕ_2 *epic is equivalent to a splitting homomorphism associated with a Heegaard splitting of some closed, orientable 3-manifold.*

PROOF. It is sufficient to produce an embedding of T as the boundary of a cube with handles which induces the map ϕ_i $(i = 1, 2)$. For this, one chooses a map of T onto a wedge, X, of g circles which induces ϕ_i.

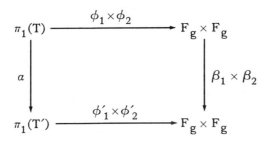

One shows that this map can be modified, by surgery, to one such that the inverse of a point in each circle is a single simple closed curve. We refer to [48] for the details of this argument.

Waldhausen [106] has shown that any two Heegaard splitting of S^3 of the same genus are equivalent. Combining this result with the above, we have

14.6 THEOREM. *The (3-dimensional) Poincaré conjecture is true if and only if for each* g *there is (up to equivalence) exactly one epimorphism*

$$\phi_1 \times \phi_2 : \pi_1(T) \to F_g \times F_g$$

of the fundamental group of a closed, orientable surface of genus g *onto the direct product of two free groups of rank* g.

PROOF. Let $(\overline{V}_1, \overline{V}_2)$ be a "standard" genus g splitting of S^3 and $\overline{\phi}_1 \times \overline{\phi}_2 : \pi_1(\overline{T}) \to \pi_1(\overline{V}_1) \times \pi_1(\overline{V}_2)$ the associated splitting epimorphism.

Let $\phi_1 \times \phi_2 : \pi_1(T) \to \pi_1(V_1) \times \pi_1(V_2)$ be the splitting epimorphism associated with a Heegaard splitting (V_1, V_2) of a simply connected 3-manifold M. By 14.4 and 14.5, any epimorphism $\pi_1(T) \to F_g \times F_g$ can be realized in this way (up to equivalence).

Suppose there is an equivalence:

$$\begin{array}{ccc}
\pi_1(\overline{T}) & \xrightarrow{\overline{\phi}_1 \times \overline{\phi}_2} & \pi_1(\overline{V}_1) \times \pi_1(\overline{V}_2) \\
\Big\downarrow \alpha & & \Big\downarrow \beta_1 \times \beta_2 \\
\pi_1(T) & \xrightarrow{\phi_1 \times \phi_2} & \pi_1(V_1) \times \pi_1(V_2)
\end{array}$$

Then, by Theorem 13.1, there is a homeomorphism $f : \overline{T} \to T$ with $f_* = \alpha$. Since $f_*(\ker \overline{\phi}_i) = \ker \phi_i$, f extends to homeomorphisms of $\overline{V}_i$ to V_i and, hence, determines a homeomorphism of S^3 to M. Conversely, if M is homeomorphic to S^3, there is, by [106], a homeomorphism $f: S^3 \to M$ such that $f(\overline{V}_i) = V_i$ $i = 1, 2$. Then f induces an equivalence of splitting epimorphisms.

The previous theorem can be reformulated as:

14.7 THEOREM. *The Poincaré conjecture is true if and only if for each* $g \geq 2$, *the kernel of each epimorphism* $\phi_1 \times \phi_2 : \pi_1(T) \to F_g \times F_g$ *contains a nontrivial element represented by a simple loop.*

PROOF. Let (V_1, V_2) be a Heegaard splitting of genus g of a closed, simply connected 3-manifold M and

$$\phi_1 \times \phi_2 : \pi_1(T) \to \pi_1(V_1) \times \pi_1(V_2)$$

the associated epimorphism. If $g = 1$, it follows easily that $M = S^3$ (cf. exercise 2.11). Suppose $g \geq 2$. If $\alpha : S^1 \to T$ is an embedding with $1 \neq [\alpha] \in \ker(\phi_1 \times \phi_2) = \ker \phi_1 \cap \ker \phi_2$, then by Dehn's lemma, $\alpha(S^1)$ bounds 2-cells D_i in V_i $(i = 1, 2)$. The 2-sphere $D_1 \cup D_2$ gives a decomposition $M = M_1 \# M_2$ where M_i has a Heegaard splitting of genus less than g $(i = 1, 2)$. It follows by induction that $M = S^3$. The converse again follows from [106].

We should note that for $\phi_1 \times \phi_2 : \pi_1(T) \to F_g \times F_g$ a splitting homomorphism associated with an *arbitrary* 3-manifold M, it is always true that $\ker(\phi_1 \times \phi_2) \neq 1$, but it does not follow that $\ker(\phi_1 \times \phi_2)$ contains an element $(\neq 1)$ represented by a simple loop; otherwise, the above proof would lead to the absurdity that every 3-manifold is a connected sum of lens spaces.

Papakyriakopoulos has established the following

14.8 THEOREM. *Let* $\phi_1 \times \phi_2 : \pi_1(T) \to F_g \times F_g$ *be the splitting homomorphism associated to a Heegaard splitting of a* 3*-manifold* M. *Then there is an embedding* $\alpha : S^1 \to T$ *with* $1 \neq [\alpha] \in \ker(\phi_1 \times \phi_2)$ *if and only if* *there is a nontrivial normal subgroup* K *of* $\pi_1(T)$ *contained in* $\ker(\phi_1 \times \phi_2)$ *with* $\pi_1(T)/K$ *torsion free and with the covering of* T *corresponding to* K *a planar surface.*

We refer to [82] for a proof (which is rather involved) of this theorem and for further comments on its applicability to the Poincaré conjecture.

The Mapping Class Group

The results of the preceding section can be reinterpreted in terms of the *mapping class group,* MC(T), of a closed orientable surface T; which is defined as the group of homeomorphisms of T modulo those which are isotopic to the identity. For $f: T \to T$ a homeomorphism, we denote by $<f> \in MC(T)$ the corresponding mapping class. We note that by 13.1 every automorphism of $\pi_1(T)$ is induced by a homeomorphism of T. Two homeomorphisms of T whose induced automorphism of $\pi_1(T)$ differ by an inner automorphism are freely homotopic, hence, [21], isotopic. Thus MC(T) is isomorphic to the group of outer automorphism of $\pi_1(T)$ (called the *homeotopy group* of T).

Let (V_1, V_2) be a Heegaard splitting of a 3-manifold M, and let $T = \partial V_1 = \partial V_2$. Choose a homeomorphism $h: T \to T$ which extends to a homeomorphism $H: V_1 \to V_2$ (h is determined up to conjugation by a homeomorphism of V_1). Observe that if $D_1, \cdots, D_g$ (g = genus T) is a set of cutting disks for V_1, then $(V_1; h(\partial D_1), \cdots, h(\partial D_g))$ is a Heegaard diagram associated to the splitting (V_1, V_2) as defined in Chapter 2. In particular, V_1 and h completely determine the splitting (V_1, V_2) of M. Let K_i be the subgroup of MC(T) consisting of those f such that f extends to a homeomorphism of V_i to itself. It follows easily from Dehn's lemma that

$$K_i = \{<f> : f_*(\ker \phi_i) = \ker \phi_i\}$$

where $\phi_i : \pi_i(T) \to \pi_i(V_i)$ is inclusion induced.

Let M′ be another 3-manifold with splitting (V'_1, V'_2), associated homeomorphism $h': T' \to T'$, etc. With this notation we have

14.9 THEOREM. *The following are equivalent*:

(i) *There is a homeomorphism* $F: M \to M'$ *such that* $F(V_i) = V'_i$.

(ii) *For some homeomorphism* $G: V_1 \to V'_1$, $ghg^{-1}h'^{-1} \in K'_1 K'_2$ *where* $g = G|T$.

(iii) *Condition* (ii) *holds for every homeomorphism* $G: V_1 \to V'_1$.

PROOF. Suppose we are given $G: V_1 \to V'_1$ with $ghg^{-1}h'^{-1} = k_1 k_2$ where $g = G|T$ and $<k_i> \in K'_i$.

Let

$$f = k_1^{-1} g = k_2 h' g h^{-1} : T \to T' .$$

From the first description f is seen to extend to a homeomorphism of V_1 to V'_1. From the second, we see that f extends to a homeomorphism of V_2 to V'_2.

If $G_1 : V_1 \to V'_1$ is any other homeomorphism and $g_1 = G_1|T$, then

$$< g_1 h g_1^{-1} h'^{-1} >$$

$$= < (g_1 g)^{-1} ghg^{-1} h'^{-1} h' (g_1 g^{-1})^{-1} h'^{-1} >$$

$$\in K'_1 < ghg^{-1}h'^{-1} > K'_2 = K'_1 K'_2 .$$

The remaining implications follow similarly.

We choose a splitting $(\overline{V}_1, \overline{V}_2)$ for S^3 (unique by [106]) where:

$$\pi_1(\overline{T}) = < a_1, b_2, \cdots, a_g, b_g : \prod [a_i, b_i] = 1> ,$$

$\ker \overline{\phi}_1$ is the normal subgroup generated by $\{b_1, \cdots, b_g\}$, $\ker \overline{\phi}_2$ is the normal subgroup generated by $\{a_1, \cdots, a_g\}$, and $\overline{h}_*(a_i) = b_i$, $\overline{h}_*(b_i) = a_i$. $(<\overline{h}^2> = 1)$

14.10 THEOREM. *The Poincaré conjecture is equivalent to the following*: *If* $k: \overline{T} \to \overline{T}$ *is a homeomorphism such that* $\ker \overline{\phi}_1 \cdot k_*(\ker \overline{\phi}_1) = \pi_1(T)$, *then* $<k\overline{h}^{-1}> \in \overline{K}_1 \overline{K}_2$.

PROOF. Let M be a 3-manifold, with splitting (V_1, V_2), associated homeomorphism $h: T \to T$, etc. as in 14.9. Note that $\pi_1(M) = 1$ if and only if $\ker \phi_1 \cdot \ker \phi_2 = \pi_1(T)$ and that $\ker \phi_2 = h_*(\ker \phi_1)$. By [106] $M = S^3$ if and only if there is a homeomorphism $F: M \to S^3$ with $F(\overline{V}_i) = V_i$ (for genus $\overline{T}$ = genus T). Let $G: V_1 \to \overline{V}_1$ be any homeomorphism and put $g = G|T$. Putting $k = ghg^{-1}$, we have $\pi_1(M) = 1$ if and only if $\ker \overline{\phi}_1 \cdot k_*(\ker \overline{\phi}_1) = \pi_1(\overline{T})$. By 14.9 $M = S^3$ if and only if $k\overline{h}^{-1} \in \overline{K}_1 \overline{K}_2$. Since any $k: \overline{T} \to T$ can be realized in this manner, we have established the equivalence.

Dehn [18] and, more recently, Lickorish [60] have explicitly described a set of generators for MC(T). This makes calculation possible and raises other obvious questions: Can one describe generators for $\overline{K}_i$? Can one restate the conditions $\ker \overline{\phi}_1 \cdot k_*(\ker \overline{\phi}_1) = \pi_1(\overline{T})$ in terms of the mapping class group, etc. For further reading on this approach we refer to Birman [9] on which our presentation is based and to [11] which describes an algorithm for determining whether a genus two Heegaard splitting represents the 3-sphere.

Involutions on Homotopy 3-Spheres

We conclude by outlining another construction which might lead to counterexamples to the Poincaré conjecture. See [39] for more details. We consider homotopy 3-spheres M which admit a free involution $\tau: M \to M$. This is prompted by the observation that $Q = M/\tau$ always contains an incompressible surface and by the fact that incompressible surfaces have thus far proved very useful in the theory. In this case, however, we will be dealing with one sided surfaces and the loop theorem no longer applies to equate incompressibility of F in Q with $\ker(\pi_1(F) \to \pi_1(Q)) = 1$. Since $M \# M$ always admits a free involution ($P^3 \# M$ is doubly covered by $M \# M$), we restrict to the case in which M is irreducible.

14.11 THEOREM. *Let* M *be an irreducible* 3-*manifold with* $|\pi_1(M)| < \infty$ *and let* $\tau: M \to M$ *be a free involution. Then* M/τ *is irreducible and contains a closed (one sided and nonorientable) incompressible surface. Furthermore, the following are equivalent:*

(i) $M = S^3$.

(ii) $M/\tau = P^3$.

(iii) M/τ *contains a projective plane.*

(iv) M/τ *does not contain a closed incompressible surface* F *with* $\chi(F) \leq 0$.

PROOF. Let $Q = M/\tau$ and $p : M \to Q$ be the covering projection.

If Q is not irreducible, then Q contains a fake 3-cell X. Each component of $p^{-1}(X)$ maps homeomorphically to X. This contradicts the irreducibility of M; for M doesn't contain any fake 3-cells unless M is a homotopy 3-sphere in which case the complement of any fake 3-cell in M is a real 3-cell.

To show that Q contains an incompressible surface, let $f : Q \to P^4$ be a map with $f_* : \pi_1(Q) \to \pi_1(P^4)$ an epimorphism. By performing surgery (Lemma 6.5 with the term 2-sided omitted), we can assume that for the standard $P^3 \subset P^4$, $f^{-1}(P^3)$ is a nonempty union of incompressible surfaces.

That (ii) $\Longrightarrow$ (i) is trivial. By a theorem of Livesay [61] that there is one equivalence class of free involutions of S^3, we have (i) $\Longrightarrow$ (ii).

Clearly (ii) $\Longrightarrow$ (iii). For (iii) $\Longrightarrow$ (ii), note that a regular neighborhood, N, of P^2 in Q is the twisted I-bundle over P^2. By irreducibility, $\overline{Q-N}$ is a 3-cell. Thus $Q = P^3$.

Since Q contains some closed, incompressible surface, it is clear that (iv) $\Longrightarrow$ (iii).

To show that (ii) $\Longrightarrow$ (iv), let F be a closed incompressible surface in P^3 in general position with $P^2 \subset P^3$. Noting that at least one of each pair of disjoint simple closed curves in P^2 bounds a 2-cell in P^2, we see that F can be isotoped so as to meet P^2 in a single curve. Since $P^3 - P^2$ is an open 3-cell, $F - P^2$ is an open 2-cell. Thus $F = P^2$.

14.12 LEMMA. *Let* M *be a closed, orientable 3-manifold,* $\tau : M \to M$ *a free involution and* $p : M \to Q = M/\tau$ *the covering projection.*

Suppose (V_1, V_2) *is a Heegaard splitting of* M *satisfying*

(i) $\tau(V_1) = V_2$, *and*

(ii) *if* $\alpha : S^1 \to \partial V_1$ *is any loop with* $1 \neq [\alpha] \in \ker(\pi_1(\partial V_1) \to \pi_1(V_1))$, *then* $\alpha(S^1) \cap \tau\alpha(S^1) \neq \emptyset$.

Then $p(\partial V_1)$ *is incompressible in* Q.

Conversely, if $|\pi_1(M)| < \infty$, *if* M *is irreducible, and if* F *is a closed, incompressible surface in* Q, *then there is a splitting* (V_1, V_2) *of* M *satisfying* (i) *and* (ii) *with* $\partial V_1 = \partial V_2 = p^{-1}(F)$.

PROOF. Given (V_1, V_2) satisfying (i) and (ii), it follows that $p(\partial V_1)$ is incompressible in Q; for if D were a compressing 2-cell of $p(\partial V_1)$, then $p^{-1}(D)$ would have two components D_1 and D_2 (with $D_i \subset V_i$). Then ∂D_1 gives a loop contradicting (ii).

For the converse, let F be closed and incompressible in Q. Observe that F must be one sided and nonorientable. A regular neighborhood, N, of F is a twisted I-bundle over F and $p^{-1}(N)$ is a product bundle over $p^{-1}(F)$. Since ∂N is two-sided in Q and $\pi_1(Q)$ is finite, ∂N is "completely compressible" in Q. Since F is incompressible, any compressing 2-cell of ∂N can be isotoped so as to lie in $\overline{Q-N}$. As in 14.11, Q is irreducible. Thus $\overline{Q-N}$ is a cube with handles. Each component of $p^{-1}(\overline{Q-N})$ maps homeomorphically to $\overline{Q-N}$. Thus $p^{-1}(F)$ bounds cubes with handles V_1, V_2 with $V_1 \cup V_2 = M$ and with $\tau(V_1) = V_2$.

Suppose we have $\alpha : S^1 \to \partial V_1$ with

$$1 \neq [\alpha] \in \ker(\pi_1(\partial V_1) \to \pi_1(V_1))$$

and with $\alpha(S^1) \cap \tau\alpha(S^1) = \emptyset$. By the loop theorem, there is a 2-cell D properly embedded in V_1 with ∂D not contractible in ∂V_1. We can require that ∂D lie in a small neighborhood of $\alpha(S^1)$. Then $D \cap \tau(D) = \emptyset$; so $p|D$ is an embedding. This contradicts the incompressibility of F.

Together, 14.11 and 14.12 provide a means of constructing examples. For each g, we fix a closed, orientable surface T_g of genus g and a free involution $\sigma : T_g \to T_g$ such that T_g/σ is a nonorientable surface with $\chi(T_g/\sigma) = 1-g$ as follows. Let T_g be embedded in R^3 in such a way that T_g is invariant under the reflection σ_1 through a plane P which intersects T_g in one or two circles according as g is even or odd and so that T_g is also invariant under a rotation σ_2 of $180°$ about a line L normal to P. Then let $\sigma = \sigma_2 \rho_1 | T_g$.

If V_1 is a cube with g handles and $h : T_g \to \partial V_1$ is a homeomorphism, then there is a 3-manifold $M(h)$ with Heegaard splitting (V_1, V_2) determined uniquely by the condition that $h\sigma h^{-1}$ extends to a free involution $\tau : M(h) \to M(h)$ with $\tau(V_1) = V_2$. If $M(h)$ satisfies condition (ii) of 14.12 and if $g \geq 1$, then it follows from 14.11 and 14.12 that $M(h) \neq S^3$. The obvious question is: under these circumstances can $M(h)$ be simply connected?

To rephrase this slightly, we say that a subgroup $N < \pi_1(T_g)$ is *self linked* if N is normal, $\pi_1(T_g)/N$ is free of rank g, and if $\alpha(S^1) \cap \sigma\alpha(S^1) \neq \emptyset$ for any loop $\alpha : S^1 \to T_g$ representing a nontrivial element of N.

Summarizing, we have

14.13 THEOREM. *There is an irreducible homotopy 3-sphere* $M \neq S^3$ *which admits a free involution if and only if for some* $g \geq 2$, *there is a self linked subgroup* $N < \pi_1(T_g)$ *such that* $N \cdot \sigma_*(N) = \pi_1(T_g)$.

PROOF. Suppose M is an irreducible homotopy 3-sphere ($\neq S^3$) and $\tau : M \to M$ is a free involution. Let $p : M \to Q = M/\tau$ be the covering projection. By 14.11, Q contains a closed (nonorientable) incompressible surface F with $\chi(F) \leq 0$. By 14.12, M has a Heegaard splitting (V_1, V_2) with $\partial V_1 = p^{-1}(F)$ and $\tau(V_1) = V_2$. Note that ∂V_1 has genus $g = 1 - \chi(F) \geq 1$. Since a simply connected lens space is S^3, we in fact have $g \geq 2$. The orientable double cover of F is unique. Thus there is a homeomorphism $h : T_g \to \partial V_1$ such that $h^{-1}\tau h = \sigma$. Let $N = \ker(h_* : \pi_1(T_g) \to \pi_1(V_1))$. By 14.12 N is self linked. Since $\pi_1(M) = 1$, $N \cdot \sigma_*(N) = \pi_1(T_g)$.

Conversely, let N be a self linked subgroup of $\pi_1(T_g)$. By Jaco's theorem [48], there is cube with g handles, V_1, and a homeomorphism $h: T_g \to \partial V_1$ with $N = \ker(h_* : \pi_1(T_g) \to \pi_1(V_1))$. If $N \cdot \sigma_*(N) = \pi_1(T_g)$, then $M(h)$ is a homotopy 3-sphere. By 14.12, $M(h)/\tau$ contains the incompressible surface $F = p(\partial V_1)$. Since $\chi(F) = 1 - g < 0$, it follows from 14.11 that $M(h) \neq S^3$. To see that $M(h)$ is irreducible, first observe that $M(h)/\tau$ is irreducible since any 2-sphere can be isotoped to miss F and hence to lift to V_1. It then follows from 10.4 that $M(h)$ is irreducible.

In the first nontrivial case, $g = 2$, the recognition of self linked subgroups is easier [39; Theorem 4.3]. We leave this as

14.14 EXERCISE. *Let* N *be a normal subgroup of* $\pi_1(T_2)$ *with* $\pi_1(T_2)/N$ *free of rank* 2. *Suppose that* $N \cap i_* \pi_1(A) = 1$ *where* A *is a component of* $T_2 - P$ (P *is the plane of symmetry of* T_2). *Show that either* N *is self linked or that* $\pi_1(T_2)/(N \cdot \sigma_*(N)$ *has infinite abelianization. Deduce that if* V_1 *is a cube with two handles and* $h: T_2 \to \partial V_1$ *is a homeomorphism with* $h(A)$ *incompressible in* V_1, *then* $M(h) \neq S^3$.

CHAPTER 15
OPEN PROBLEMS

The ultimate goal of the theory would be in providing solutions to:

The *homeomorphism problem*: provide an "effective" procedure for determining whether two "given" 3-manifolds are homeomorphic, together with

The *classification problem*: "effectively" generate a list containing exactly one 3-manifold from each homeomorphism class.

Of course the terms "effective" and "given" must be used in a precise way and the second allows many interpretations. However, it is generally agreed that under reasonable interpretations, the corresponding problems for 2-manifolds are solved and for n-manifolds, $n \geq 4$, are unsolvable [12], [65].

It remains open whether these problems are solvable for 3-manifolds and if so, to provide efficient solutions. We will suggest some steps toward these problems and raise a number of related questions. We will only be concerned with compact 3-manifolds.

The Fundamental Groups

Our emphasis in this work has been the role of the fundamental group (the fundamental group system in the case of bounded 3-manifolds) in determining the structure of the corresponding 3-manifold. Even where the fundamental group is known to be insufficient to characterize the manifold (e.g., lens spaces), the distinguishing invariant (e.g., Reidemeister torsion) depends on the fundamental group (its action on the homology groups). It is conceivable that, assuming the Poincaré conjecture is true, all invariants depend on the fundamental group. Conversely, if the homeomorphism, classification, or any other decision problems regarding 3-manifolds turn out to be unsolvable, it is likely that this will be shown by reduction to a problem about the corresponding class of fundamental groups. Because of prime factorization (3.15, 3.21) and Kneser's conjecture (7.1), it is reasonable to consider *the class of fundamental groups of compact, prime, 3-manifolds* which we denote by $\mathcal{G}$. The subclasses corresponding to orientable, nonorientable, sufficiently large, closed, and bounded 3-manifolds will be denoted respectively by $\mathcal{G}_O$, $\mathcal{G}_{NO}$, $\mathcal{G}_{SL}$, $\mathcal{G}_C$, and $\mathcal{G}_B$.

Our general goal is to gain information about $\mathcal{G}$. The following summarizes some of the known information.

15.1. (a) *The finite groups in* $\mathcal{G}$ *are included in the list of Milnor* [69] *as restricted by Lee* [58]; Z_2 *is the only finite group in* $\mathcal{G}_{NO}$ (9.5).

(b) *The infinite groups in* $\mathcal{G}_0$ *are torsion free and in* $\mathcal{G}_{NO}$ *have at most* 2*-torsion* (9.8).

(c) *The abelian groups in* $\mathcal{G}$ *are given by* (9.13). *The nilpotent groups in* $\mathcal{G}$ *are described by Thomas* [102], *the solvable groups in* $\mathcal{G}$ *are described by Evans and Moser* [23].

(d) *The torsion free groups in* $\mathcal{G}$ *have homological dimension* ≤ 3 (h.d. ≤ 2 *in* $\mathcal{G}_B$) (4.3 *and* 4.12).

(e) $\mathcal{G}$ *is closed under the operation of taking finitely generated subgroups* (8.2).

(f) *The extensions* $1 \to N \to G \to Q \to 1$ *with* $G \in \mathcal{G}$, N *finitely generated, and* $|Q| = \infty$ *are described in Chapters* 11 *and* 12.

(g) *For* $G \in \mathcal{G}$, *the deficiency,*

$$\operatorname{def} G = \sup\{n-m : G = \langle x_1, \cdots, x_n : r_1, \cdots, r_m \rangle\} ,$$

can be explicitly expressed in terms of a corresponding 3-manifold [19], [52]. *For example, for* $G \in \mathcal{G}_O$, $\operatorname{def} G \geq 0$; *for* $Z \neq G \in \mathcal{G}_C \cap \mathcal{G}_O$; $\operatorname{def} G = 0$.

Of central importance is

15.2 PROBLEM. *Characterize the class* $\mathcal{G}$ *(or any of the subclasses).*

Such a problem may, of course, allow many different sorts of solutions and in fact different interpretations. Some may be impossible: for example, that of finding an algorithm for determining whether arbitrary finite presentations present groups in $\mathcal{G}$. Observe that $G \times Z \times Z \times Z$ is the fundamental group of *any* 3-manifold if and only if $G = 1$. Thus such an algorithm would solve the triviality problem for the class of all finitely presented groups.

Observe that if M is a prime 3-manifold, then M is P^2-irreducible unless $\pi_1(M)$ has 2-torsion (and M is nonorientable) or M is a 2-sphere bundle over S^1 (in which case, as far as fundamental groups are concerned, we can replace M by $B^2 \times S^1$). If M is P^2-irreducible, then its universal cover, $\tilde{M}$, is a homotopy 3-sphere if $\pi_1(M)$ is finite or a contractible 3-manifold if $\pi_1(M)$ is infinite.

We will say that a compact 3-manifold is a *sphere-quotient* or *cell-quotient* provided that its universal cover is homeomorphic to either S^3 or to $B^3 - K$ for some compact set $K \subset \partial B^3$. By (13.4), every compact, P^2-irreducible, sufficiently large 3-manifold is a cell-quotient. Whether every compact, P^2-irreducible 3-manifold is a sphere or cell-quotient depends on the Poincaré conjecture and, independently on

15.3 QUESTION. *Let* M *be a noncompact, contractible, irreducible 3-manifold with* $\mathrm{Int}\, M \neq R^3$ (e.g., *example* 14.1). *Can there exist a properly discontinuous group of homeomorphisms of* M *with compact quotient?*

As far as we know, the following is not known.

15.4 QUESTION. *Is every compact* P^2*-irreducible 3-manifold homotopy equivalent to a sphere or cell-quotient?*

Since $\pi_1(M)$ is the group of covering transformations of the universal cover of M, it is natural to ask

15.5 QUESTION. *If* $G \in \mathcal{G}$ *is finite, is* G *isomorphic to a subgroup of* SO(4)?

15.6 QUESTION. *If* $G \in \mathcal{G}$ *is infinite and torsion free, is* G *isomorphic to some properly discontinuous group of birational homeomorphisms of some contractible submanifold of* R^3? *(A homeomorphism* h *is birational if the coordinate functions of* h *and of* h^{-1} *are rational functions.)*

Affirmative answers would provide useful algebraic information. In the case of sphere or cell-quotients, we can ask the more demanding

15.7 QUESTION. *If* M *is a cell-quotient (sphere-quotient) with universal cover* $\tilde{M}$, *is there an embedding of* $\tilde{M}$ *in* R^3 *(a homeomorphism of* $\tilde{M}$ *with* S^3*) which identifies the group of covering transformations with a group of birational homeomorphisms (a subgroup of* $SO(4)$*)?*

Little is known about 15.7. Even in the case in which $\pi_1(M)$ is finite cyclic affirmative answers are known only if $\pi_1(M)$ has order two [61], four [86], or eight [87].

Peripheral Systems

For $G \in \mathcal{G}$, a collection $\{H_1, \cdots, H_k\}$ of subgroups of G is called a *peripheral system* in G if there is a compact 3-manifold M, pairwise disjoint, incompressible surfaces $F_1, \cdots, F_k$ in ∂M, paths τ_i joining a base point x_0 of M to a base point x_i of F_i, and an isomorphism $\phi: G \to \pi_1(M, x_0)$ such that $\phi(H_i) = h_{\tau_i} i_* \pi_1(F_i, x_i)$ (h_{τ_i} the change of base point isomorphism determined by τ_i). The H_i are called *peripheral subgroups*. If F_i is a component of ∂M, H_i will be called a *peripheral component*. Peripheral systems $\{H_1, \cdots, H_k\}$ in G and $\{H'_1, \cdots, H'_k\}$ in G' are *equivalent* if there is an isomorphism $\psi: G \to G'$ such that $\psi(H_i)$ is conjugate to $H'_{j(i)}$ for some permutation $j(1), \cdots, j(k)$ of $1, \cdots, k$.

15.8 QUESTION. *How can the peripheral subgroups (systems) be detected among all subgroups (systems of subgroups) of a group* $G \in \mathcal{G}$?

Corresponding to nonhomeomorphic, sufficiently large 3-manifolds with isomorphic fundamental groups (cf. 13.11), we have groups $G \in \mathcal{G}$ with inequivalent systems of peripheral components.

15.9 QUESTION. *Can a group* $G \in \mathcal{G}$ *have infinitely many inequivalent systems of peripheral components?*

An understanding of the peripheral structure in applying the results of Chapter 13 (13.6, 13.7, 13.9) is of obvious importance. It may be of greater importance in deducing properties of the class $\mathcal{G}_{SL}$. Since every P^2-irreducible, sufficiently large 3-manifold can be reduced to a 3-cell (or a union of two 3-cells) by a sequence of cuts along properly embedded, 2-sided incompressible surfaces (13.3), we have a procedure for building up its fundamental group by a sequence of HNN constructions and (possibly one) amalgamated free product. Specifically, let M' be obtained by cutting M along an incompressible surface F. Suppose M' is connected. We have monomorphisms $\eta_i : \pi_1(F) \to \pi_1(M)$ $(i = 0, 1)$ induced by the obvious embeddings of F in $\partial M'$ and reference to a common base point. Then $\pi_1(M)$ is a *Higman-Neumann-Neumann* (HNN) *group* with *base* $\pi_1(M')$ and *bonding subgroups* $\eta_i(\pi_1(F))$ $i = 0, 1$ given explicitly by

$$\pi_1(M) = (\pi_1(M') * \langle t : - \rangle)/N$$

where N is the smallest normal subgroup containing

$$\{t\eta_0(a)t^{-1}\eta_1(a)^{-1};\ a \in \pi_1(F)\} .$$

In general, given a group B, subgroups C_0 and C_1 and an isomorphism $\phi : C_0 \to C_1$, the group

$$G = (B * \langle t : - \rangle)/N$$

where N is the smallest normal subgroup containing

$$\{tct^{-1}\phi(c)^{-1} : c \in C_0\}$$

is called the HNN *group* with *base* B, *bonding subgroups* C_0 and C_1, and *bonding map* ϕ, and is denoted $G = \mathrm{HNN}(B, C_0, C_1, \phi)$.

If M' has two components M'_0 and M'_1, we have an analogous representation

$$\pi_1(M) = \pi_1(M'_0) \underset{\eta_0\pi_1(F)=\eta_1\pi_1(F)}{*} \pi_1(M'_1)$$

of $\pi_1(M)$ as a *free product amalgamated* along the subgroups $\eta_i(\pi_1(F))$.

While it is known (cf. example 15.15 below) that quite general (and pathological) groups can be built up from "nice" (e.g., free) groups as HNN groups and/or amalgamated free products by using arbitrary (finitely generated) bonding subgroups, the restrictions given by 15.1 indicate that this is not the case with 3-manifolds. We feel that this may be explained by the fact that, in the case of 3-manifolds, the bonding subgroups form a peripheral system, and would like to understand what "placement properties" of peripheral systems can be used in studying the class $\mathcal{G}_{SL}$ as a class of HNN groups.

As an example of a nice placement property of a peripheral subgroup, we prove

15.10 LEMMA. *If* $G \in \mathcal{G}$ *and* $H < G$ *is peripheral, then for any finitely generated* $K < G$, $H \cap K$ *is finitely generated.*

PROOF. There is a compact 3-manifold M with an incompressible surface $F \subset \partial M$ such that $(\pi_1(M), \pi_1(F)) \cong (G, H)$. Let (M', F') be a disjoint copy of the pair (M, F) and form $M_1 = M \underset{F=F'}{\cup} M'$. Then $\pi_1(M_1) \cong G \underset{H=H'}{*} G'$. If for some finitely generated $K < G$, $K \cap H$ is not finitely generated, then (cf. [3]) the subgroup of $\pi_1(M_1)$ generated by $K \cup K'$ is seen to be finitely generated, but not finitely related. This contradicts (8.2).

To see, by contrast, that two arbitrary finitely generated subgroups of a group $G \in \mathcal{G}$ need not have finitely generated intersection, let F be a closed orientable surface of genus $g \geq 2$. Then, by 11.12, for any $n \geq 1$, $F \times S^1$ also fibers over S^1 with fiber a closed orientable surface F_1 of genus $n(g-1)+1$.

15.11 EXERCISE. *Show that* $\pi_1(F) \cap \pi_1(F_1)$ *is not finitely generated. Hint: if* $\pi_1(F) \cap \pi_1(F_1)$ *is finitely generated, being normal, it has finite index in each of* $\pi_1(F), \pi_1(F_1)$.

Hopficity

In a sense, the results of Waldhausen (Chapter 13) solve the homeomorphism problem for P^2-irreducible, sufficiently large 3-manifolds. The solution, however, cannot be considered effective; since it reduces to the isomorphism problem for the corresponding fundamental groups (group systems in the bounded case). In an attempt to reduce this to a more tractable problem, we note

15.12 LEMMA. *If* $f:(M, \partial M) \to (N, \partial N)$ *is a degree one map of compact oriented manifolds of the same dimension (with possibly empty boundary), then* $f_*: \pi_1(M) \to \pi_1(N)$ *is an epimorphism.*

PROOF. There is a covering $p: \tilde{N} \to N$ with $p_* \pi_1(\tilde{N}) = f_*(\pi_1(M))$ and a factoring $\tilde{f}: M \to N$ with $p \circ \tilde{f} = f$. Now $\deg f = \deg \tilde{f} \cdot \deg p$, but $\deg p = 0$ if $\tilde{N}$ is noncompact and $\deg p = [\pi_1(N) : p_* \pi_1(\tilde{N})]$ if N is compact. Thus p_*, hence f_*, is epic.

A group G is said to be *Hopfian* if each epimorphism of G to itself is an isomorphism. Combining 15.12 with 13.7, we have

15.13 THEOREM. *Suppose* M *and* N *are compact, orientable,* P^2*-irreducible, sufficiently large 3-manifolds with Hopfian fundamental groups. Then* M *is homeomorphic to* N *if and only if there exist degree one maps* $f:(M, \partial M) \rightleftarrows (N, \partial N) : g$.

15.14 QUESTION. *Is each* $G \in \mathcal{G}$ *(or* $G \in \mathcal{G}_{SL}$*) Hopfian? Or (what we really need), if* $G \in \mathcal{G}$ *is each epimorphism of* G *to itself which preserves some fixed system of peripheral components an isomorphism?*

The simplest examples of finitely presented, non-Hopfian groups are given by

15.15 EXAMPLE [4]. *For* $(p,q) = 1$, *the group* $G_{p,q} = \langle a,b : a^{-1}b^pa = b^q\rangle$ *is non-Hopfian.*

The map $\psi(a) = x$, $\psi(b) = zw^{-1}$ *induces an embedding of* $G_{p,q}$ *into*

$$H_{p,q} = \langle x,y : -\rangle \underset{\substack{x=z \\ y^{-1}x^py = w^{-1}z^qw}}{*} \langle z,w : -\rangle$$

which is a free product of two rank two free groups amalgamated along rank two subgroups.

PROOF. The map $\phi(a) = a$, $\phi(b) = b^p$ preserves relations and hence determines a homomorphism whose image contains a, b^p, $a^{-1}b^pa = b^q$, and hence b (since $(p,q) = 1$). One must check that

$$1 \neq (a^{-1}bab^{-1})^p b^{p-q} \in \ker\phi \ .$$

We leave the second part as an exercise.

Observe that $G_{p,q}$ is an HNN group with infinite cyclic base. However, ad hoc arguments exist [37] for showing that $G_{p,q} \notin \mathcal{G}$.

Residual Finiteness

A group G is said to be *residually finite* if for each $1 \neq g \in G$, there is a finite group H and a homomorphism $\phi : G \to H$ with $\phi(g) \neq 1$. See [62] for further reading.

15.16 LEMMA.

(i) *If* G *is residually finite, then so is every subgroup of* G.

(ii) *If* G *is finitely generated, then* G *is residually finite if and only if* $\cap\{H < G : [G:H] < \infty\} = 1$, *in particular*

(iii) *If* G *is finitely generated and* G *contains a residually finite subgroup of finite index, then* G *is residually finite.*

PROOF. Part (i) is clear.

For (ii), note that for each integer n, G contains only finitely many subgroups of index n. This follows in the case G is free by enumerating all n-sheeted coverings of a finite wedge of circles and hence for arbitrarily finitely generated G by taking quotients. Thus for each i, $\theta_i(G) = \cap\{H < G : [G:H] \leq i\}$ is a normal (fully invariant) subgroup of finite index in G. Parts (ii) and (iii) follow easily from this.

Our interest in this property is

15.17 LEMMA [64]. *Each finitely generated, residually finite group is Hopfian.*

PROOF. Suppose G is finitely generated and $\phi : G \to G$ is an epimorphism. Then for each n, ϕ induces a one-to-one correspondence between the set of subgroups of G of index n and the set of subgroups of G of index n which contain $\ker \phi$. Since G is finitely generated, these sets are finite. Thus $\ker \phi$ lies in every subgroup of finite index in G. Since G is residually finite, $\ker \phi = 1$.

15.18 QUESTION. *Is each* $G \in \mathcal{G}$ *(or* $G \in \mathcal{G}_{SL}$*) residually finite?*

It is known that fundamental groups of 2-manifolds (including free groups) are residually finite (see [38] for a simple proof). Hence, by 12.2 and 15.16, Fuchsian groups are residually finite. From this, it follows easily that certain 3-manifolds (e.g., bundles and Seifert fibered spaces) have residually finite fundamental groups.

15.19 THEOREM. *Suppose* M *is a compact* 3*-manifold and that* $\pi_1(M)$ *contains a finitely generated normal subgroup,* $N \neq 1$, *of infinite index. If* $N \cong Z$, *assume* M *is sufficiently large. Then* $\pi_1(M)$ *is residually finite.*

PROOF. We assume $M = \mathcal{P}(M)$. Using 11.1 or 12.8, together with 12.2, it follows that either $\pi_1(M) = Z + Z_2$ or some finite sheeted cover, $\tilde{M}$, of

M (corresponding to the inverse of an infinite cycle or surface subgroup of finite index in $\pi_1(M)/N$) is either a surface bundle over S^1 or an S^1 bundle over a surface. It follows from 15.21 below in the first case and by a direct computation in the second that $\pi_1(\tilde{M})$, hence $\pi_1(M)$, is residually finite.

Residual finiteness of certain classes of knot groups has been established [67], [100] by algebraic methods.

One approach to establishing residual finiteness for the class $\mathcal{G}_{SL}$ is to consider these groups as HNN groups. Example 15.15 shows that we must proceed with caution. In the case of 3-manifolds, the bounding subgroups form a peripheral system in the base group. Since we believe that peripheral systems are nicely placed, we look for conditions on the placement of the bonding subgroups which will insure that an HNN group with residually finite base is residually finite. We illustrate some examples.

15.20 LEMMA. *Let* $G = \mathrm{HNN}(B, C_0, C_1, \phi)$ *with* B *and* C_i *finitely generated. Suppose there is a sequence* $\{N_i\}$ *of normal subgroups of finite index in* B *satisfying*

(i) $\cap N_i = 1$,

(ii) $\cap N_i C_0 = C_0$, $\cap N_i C_1 = C_1$, *and*

(iii) $\phi(N_i \cap C_0) = N_i \cap C_1$ *for all* i.

Then G *is residually finite.*

PROOF. Using mapping cylinders, we can construct a complex X with disjoint subcomplexes Y_0, Y_1 having collar neighborhoods and a homeomorphism $f: Y_0 \to Y_1$ such that for $y_0 \in Y_0$, $y_1 = f(y_0)$ and some path $\tau: I \to X$ with $\tau(0) = y_0$, $\tau(1) = y_1$, we have $B = \pi_1(X, y_0)$, $i_{0*}: \pi_1(Y_0, y_0) \to \pi_1(X, y_0)$ monic with image C_0, $i_{1*}: \pi_1(Y_1, y_1) \to \pi_1(X, y_1)$ monic with $C_1 = h_\tau(i_{1*}\pi_1(Y_1, y_1))$, and with $\phi = h_\tau \circ f_*$. Here $h_\tau: \pi_1(X, y_1) \to \pi_1(X, y_0)$ is the change of base point isomorphism corresponding to τ. We let Z be the space obtained from X by identifying each $y \in Y_0$ with $f(y) \in Y_1$

and let $\eta: X \to Z$ be the identification map. Let $Y = \eta(Y_0) = \eta(Y_1)$ and $z_0 = \eta(y_0) = \eta(y_1)$. We have a natural identification $G = \pi_1(Z, z_0)$.

A loop $\alpha: (I, \partial I) \to (Z, z_0)$ is *elementary* if $\alpha(\text{Int } I) \cap Y = \emptyset$. Any $g \in G$ can be represented by a product $\alpha = \alpha_1 \cdot \alpha_2 \cdot \cdots \cdot \alpha_n$ of elementary loops. The number of loops in a minimal representation is denoted by $\lambda(g)$. If $\lambda(g) \geq 2$, then in any minimal representation $[\alpha_i] \notin \pi_1(Y, z_0)$ for all i.

We will prove the stronger condition:

(*) If $1 \neq g \in G$ and $\alpha = \alpha_1 \cdot \cdots \cdot \alpha_{\lambda(g)}$ is a minimal representation of g, then there is a finite sheeted regular covering $p: \tilde{Z} \to Z$ such that no lifting of α is a loop. Except in the case $\lambda(g) = 1$ and $g \in \pi_1(Y, z_0)$, the liftings of α will begin and end in different components of $p^{-1}(Y)$.

The proof is by induction on $\lambda(g)$. If $\lambda(g) = 1$, there is a path $\beta: I \to X$ with $\eta \circ \beta = \alpha$ and either (a) β a loop based at y_0, (b) β a loop based at y_1, or (c) β a path joining y_0 and y_1.

In case (a) there is, by assumption (i) a normal subgroup N of finite index in $B = \pi_1(X, y_0)$ with $[\beta] \notin N$ and with $\phi(N \cap C_0) = N \cap C_1$. If $[\alpha] \notin \pi_1(Y, z_0)$, then $[\beta] \notin \pi_1(Y_0, y_0)$; so by (ii) we may assume that $[\beta] \notin NC_0$. Let $q: \overline{X} \to X$ be the finite sheeted, regular covering with $q_*\pi_1(\overline{X}) = N$. Then β does not lift to a loop in $\overline{X}$. Furthermore, if $[\beta] \notin NC_0$, $\beta(0)$ and $\beta(1)$ lie in different components of $q^{-1}(Y_0)$. Let $\overline{Y}_0$ be a component of $q^{-1}(Y_0)$, $\overline{y}_0 \in \overline{Y}_0 \cap q^{-1}(y_0)$, and $\overline{\tau}$ be the lifting of τ beginning at $\overline{y}_0$. Let $\overline{Y}_1$ be the component of $q^{-1}(Y_1)$ containing $\overline{y}_1 = \overline{\tau}(1)$. Now $q_*\pi_1(\overline{Y}_0, \overline{y}_0) = N \cap C_0$ and $q_*\pi_1(\overline{Y}_1, \overline{y}_1) = h_\tau^{-1}(N \cap C_1)$. Since $f_* = h_\tau^{-1} \circ \phi$ and $\phi(N \cap C_0) = N \cap C_1$, it follows that $f_*q_*\pi_1(\overline{Y}_0, \overline{y}_0) = q_*\pi_1(\overline{Y}_1, \overline{y}_1)$. Thus there is a homeomorphism $\overline{f}: \overline{Y}_0 \to \overline{Y}_1$ such that $f \circ q|\overline{Y}_0 = q \circ \overline{f}$. If $\overline{Y}'_0$ and $\overline{Y}'_1$ are other components of $q^{-1}(Y_0)$ and $q^{-1}(Y_1)$ respectively, choose covering translations σ_0, σ_1 such that $\sigma_i(\overline{Y}_i) = \overline{Y}'_i$. Then $\overline{f}' = \sigma_1 \circ \overline{f} \circ \sigma_0^{-1}: \overline{Y}'_0 \to \overline{Y}'_1$ satisfies $f \circ q|\overline{Y}'_0 = q \circ \overline{f}'$. Since $q^{-1}(Y_0)$ and $q^{-1}(Y_1)$ have the same number of components, we can use $\overline{f}, \overline{f}'$, etc., to construct a commutative diagram

$$\begin{array}{ccc} \bar{X} & \xrightarrow{\ \bar{\eta}\ } & \bar{Z} \\ \Big\downarrow q & & \Big\downarrow \bar{p} \\ X & \xrightarrow{\ \eta\ } & Z \end{array}$$

where $\bar{\eta}$ is an identification map and $\bar{p}$ is a covering. Now $\bar{p}$ satisfies the conditions of (*) except that it may not be regular. However, the regular cover $\tilde{p}: \tilde{Z} \to Z$ corresponding to the normal subgroup obtained by intersecting $\bar{p}_* \pi_1(\bar{Z})$ with its finitely many distinct conjugates does satisfy (*).

In case (b), we use the same construction except that we choose N so that $[\tau \cdot \beta \cdot \tau^{-1}] \notin N$ and if $[\alpha] \notin \pi_1(Y)$, $[\tau \cdot \beta \cdot \tau^{-1}] \notin NC_1$ (and so that $\phi(N \cap C_0) = N \cap C_1$).

Case (c) is the easiest; for in this case we may take $p: \tilde{Z} \to Z$ to be the 2-sheeted cover corresponding to the smallest normal subgroup of G containing B and t^2.

Now suppose $n > 1$ and that (*) is satisfied for all $h \in G$ with $\lambda(h) < n$. Let g have a minimal representation $\alpha = \alpha_1, \cdots, \alpha_n$. Recall that $\alpha_i \notin \pi_1(Y)$ for all i. By repeated applications of induction and by intersecting the corresponding normal subgroups, we obtain a finite sheeted regular covering $\bar{p}: \bar{Z} \to Z$ such that the liftings of each subproduct $\alpha_i \alpha_{i+1} \cdots \alpha_{i+j}$ of length less than n begin and end in different components of $\bar{p}^{-1}(Y)$. Let $\bar{\alpha}$ be a lifting of α. We are done unless $\bar{\alpha}(0)$ and $\bar{\alpha}(1)$ lie in the same component $\bar{Y}$ of $\bar{p}^{-1}(Y)$. If this occurs, then $\bar{\alpha}$ must meet some other component $\bar{Y}'$ of $\bar{p}^{-1}(Y)$ and do so in exactly one point. As in case (c) $(n=1)$. There is a double cover $\bar{\bar{p}}: \bar{\bar{Z}} \to \bar{Z}$ such that a lifting $\bar{\bar{\alpha}}$ of $\bar{\alpha}$ begins and ends in different components of $\bar{\bar{p}}^{-1}(\bar{Y})$. Then $\bar{p} \circ \bar{\bar{p}}: \bar{\bar{Z}} \to Z$ satisfies (*) except for being regular. As before, we choose $\tilde{p}: \tilde{Z} \to Z$ to be a finite sheeted regular cover which factors through $\bar{\bar{Z}}$.

15.21 COROLLARY [45]. *Cyclic extensions of finitely generated residually finite groups are residually finite.*

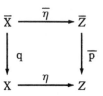

PROOF. This is clear for a finite cyclic extension.

If G contains a normal subgroup B with G/B infinite cyclic, then $G = HNN(B, B, B, \phi)$ for some automorphism $\phi : B \to B$. If B is finitely generated and residually finite, then the conditions of 15.20 are satisfied for any cofinal sequence $\{N_i\}$ of fully invariant subgroups of finite index in B (e.g., $N_i = \theta_i(B) = \cap\{H < B : [B:H] \leq i\}$).

In reconstructing a sufficiently large 3-manifold from a hierarchy, the first nontrivial stage occurs when one identifies a pair of nonsimply connected, incompressible surfaces in the boundary of a cube with handles. The corresponding fundamental group is an HNN group with free base. In this case, we can say a little more about residual finiteness. The following concept is convenient. If H is a subgroup of a group G, a *completion of* H *in* G is a subgroup K of finite index in G which contains H as a free factor; i.e., $K = H * H'$ for some subgroup H' of G.

15.22 LEMMA [35], [54]. *Each finitely generated subgroup of a free group* G *has a completion in* G.

PROOF. Let H be a finitely generated subgroup of G (which we may also assume is finitely generated). Let $G = \pi_1(\Gamma, x_0)$ where Γ is a wedge of circles and let $p : \tilde{\Gamma} \to \Gamma$ be the covering such that $p_* \pi_1(\tilde{\Gamma}, \tilde{x}_0) = H$. Choose a finite subcomplex K of $\tilde{\Gamma}$ containing x_0 minimal with respect to the property that $i_* : \pi_1(K, \tilde{x}_0) \to \pi_1(\tilde{\Gamma}, \tilde{x}_0)$ is an isomorphism. Without introducing additional vertices, one can embed K in a finite 1-complex $\bar{\Gamma}$ in such a way that $p|K$ extends to a covering map $\bar{p} : \bar{\Gamma} \to \Gamma$. Then $\bar{p}_* \pi_1(\bar{\Gamma}, \tilde{x}_0)$ is a completion of H in G.

15.23 LEMMA. *Let* $G = HNN(B, C_0, C_1, \phi)$ *where* B *is free and* C_i *is finitely generated.*

Suppose

(i) *Each of* C_0, C_1 *is a free factor of the subgroup* D *generated by* $C_0 \cup C_1$, *and*

(ii) D *has a normal completion in* B.

Then G *is residually finite.*

PROOF. Observe that hypothesis (i) is satisfied in the extreme cases $C_0 = C_1$ and $D = C_0 * C_1$.

For any finitely generated group A, let $\theta_i(A) = \{K < A : [A:K] \leq i\}$. Observe that $\theta_i(A)$ is a fully invariant subgroup of finite index in A and that A is residually finite if and only if $\cap \theta_i(A) = 1$.

Now for the proof of 15.23, we may assume that B is finitely generated and that D has a normal completion, H, in B. Let $N_i = \theta_i(H)$. Then N_i is fully invariant in H, hence normal in B. It follows from Lemma 15.24 below that $\{N_i\}$ satisfies the condition of Lemma 15.20 (e.g., $N_i \cap C_j = \theta_i(C_j)$; so $\phi(N_i \cap C_0) = N_i \cap C_1$), and the conclusion follows.

15.24 LEMMA. *Let* A *and* B *be finitely generated groups. Then*

(i) $\theta_i(A * B) \cap B = \theta_i(B)$ *for all* i, *and*

(ii) *if* A *and* B *are residually finite,* $\bigcap_i (\theta_i(A * B) B) = B$.

PROOF. Let (Z, z_0) be the wedge of complexes (X, z_0), (Y, z_0) with fundamental groups A and B respectively.

For (i), the inclusion $\theta_i(B) < \theta_i(A * B) \cap B$ is immediate. For the opposite inclusion, suppose $\beta : (I, \partial I) \to (B, z_0)$ represents $[\beta] \notin \theta_i(B)$. Then there is a covering $p : \tilde{Y} \to Y$, with at most i-sheets such that some lifting, $\tilde{\beta}$, of β is not a loop. We can extend p to a cover $q : \tilde{Z} \to Z$ (same number of sheets) by attaching a copy of X at each point of $p^{-1}(z_0)$. Then $[\beta] \notin q_*(\pi_1(\tilde{Z}, \tilde{\beta}(0))$ and $[A * B : q_*\pi_1(\tilde{Z}, \tilde{\beta}(0)] \leq i$; hence $[\beta] \notin \theta_i(A*B)$.

For (ii) we proceed as in the proof of 15.20. It suffices to show that if $\alpha : (I, \partial I) \to (Z, z_0)$ represents $[\alpha] \notin B$, then there is a finite sheeted cover $p : \tilde{Z} \to Z$ such that some lifting $\tilde{\alpha}$ of α begins and ends in different components of $p^{-1}(Y)$. We represent $\alpha = \alpha_1 \cdots \alpha_n$ minimally as a product of paths α_i which lie in one of X or Y, and induct on the length n of α. For $n = 1$, $[\alpha] \in A$ and we construct $\tilde{Z}$ by attaching copies of Y to an appropriate finite sheeted cover of X (obtained by the assumption that A is residually finite). If $n \geq 2$ we can, by induction, obtain a finite sheeted, regular cover $\bar{p}: \bar{Z} \to Z$ such that if $\beta = \alpha_i \alpha_{i+1} \cdots \alpha_j$ is any subproduct of length less than n, with $[\beta] \notin B$, then no lifting of β begins and ends in the same component of $\bar{p}^{-1}(Y)$. We may also assume that no lifting of any α_i is a loop. Let $\bar{\alpha}$ be a lifting of α to $\bar{Z}$. We are done unless $\bar{\alpha}(0)$ and $\bar{\alpha}(1)$ lie in the same component $\bar{Y}$ of $\bar{p}^{-1}(Y)$. In this case, it follows that $\bar{\alpha}(I)$ meets some other component $\bar{Y}'$ of $\bar{p}^{-1}(Y)$ and crosses some point $\bar{z}_0$ of $p^{-1}(z_0) \cap \bar{Y}'$ exactly once. Then we can choose $\tilde{Z}$ to be a double cover of $\bar{Z}$ as in the proof of 15.20.

The techniques of 15.23 can be modified in numerous ways, for example

15.25 LEMMA. *Let* $G = \mathrm{HNN}(B, C_0, C_1, \phi)$ *where* B *is free and* C_i *is finitely generated. Suppose there is a normal subgroup* K *of finite index in* B *which lies in some completion of* C_0 *and in some completion of* C_1 *and such that* $\phi(K \cap C_0) = K \cap C_1$. *Then* G *is residually finite.*

PROOF. We have completions D_j of C_j such that $K < D_0 \cap D_1$. By the Kurosh subgroup theorem, $K = K \cap D_j$ is a completion of $K \cap C_j$. Let $N_i = \theta_i(K)$. Since K is normal in B, N_i is normal and of finite index in B. By 15.24, $N_i \cap C_j = N_i \cap (K \cap C_j) = \theta_i(K \cap C_j)$. Since $\phi(K \cap C_0) = K \cap C_1$, we have $\phi(N_i \cap C_0) = N_i \cap C_1$. Finally, $N_i C_j < \theta_i(D_j) C_j$ and by 15.24, $\bigcap_i \theta_i(D_j) C_j = C_j$; so $\bigcap_i N_i C_j = C_j$. Thus $\{N_i\}$ satisfies the conditions of 15.20 and the conclusion follows.

15.26 COROLLARY. *Let* $G = \mathrm{HNN}(B, C_0, C_1, \phi)$ *where* B *is free and* C_j *is finitely generated. If* ϕ *extends to an automorphism of* B, *then* G *is residually finite.*

PROOF. We may assume that B is finitely generated. Let D_j be a completion of C_j and let $K = \theta_i(B)$ where $i = [B : D_0 \cap D_1]$. Since ϕ extends to an automorphism Φ of B and $\Phi(K) = K$, we have $\phi(K \cap C_0) = K \cap C_1$. Thus K satisfies the conditions of 15.24 and the conclusion follows.

While it is not difficult to find applications of the above lemmas to specific classes of 3-manifolds, it remains to be seen whether they will lead to any general results. Further progress in this direction will depend on our ability to recognize peripheral systems and to deal with problems such as

15.27 QUESTION. *Suppose* $B \in \mathcal{G}_{SL}$ *and* $\{C_0, C_1\}$ *is a peripheral system in* B. *What comparisons can be made between the system of subgroups of finite index in* B *with the corresponding systems in* C_0 *and* C_1? *For example, need there be any relationship between the sequences* $\{\theta_i(B) \cap C_j\}$ *and* $\{\theta_i(C_j)\}$? *If not, can these sequences be replaced by some other canonically defined sequence of (fully invariant) subgroups of finite index for which some useful answer exists?*

REFERENCES

[1] J. W. Alexander, "Note on Riemann spaces," Bull. Amer. Math. Soc. 26 (1920), 370-372.

[2] ______________, "The combinatorial theory of complexes," Ann. of Math. 31 (1930), 292-320.

[3] Gilbert Baumslag, "A remark on generalized free products," Proc. Amer. Math. Soc. 13 (1962), 53-54.

[4] Gilbert Baumslag and D. Solitar, "Some two-generator, one relator non-Hopfian groups," Bull. Amer. Math. Soc. 68 (1962), 199-201.

[5] Gilbert Baumslag and T. Taylor, "The center of groups with one defining relator," Math. Ann. 175 (1968), 315-319.

[6] R. H. Bing, "An alternative proof that 3-manifolds can be triangulated," Ann. of Math., 69 (1959), 37-65.

[7] __________, "Necessary and sufficient conditions that a 3-manifold be S^3," Ann. of Math., 68 (1958), 17-37.

[8] R. H. Bing and J. M. Martin, "Cubes with knotted holes," Trans. Amer. Math. Soc. 155 (1971), 217-231.

[9] Joan S. Birman, "Poincaré's conjecture and the homeotopy group of a closed, orientable, 2-manifold," J. Aust. Math. Soc. XVII (1974), 214-221.

[10] ______________, *Braids, Links, and Mapping Class Groups*, Ann. of Math Studies No. 82, Princeton University Press (1975).

[11] Joan S. Birman and Hugh W. Hilden, "The homeomorphism problem for S^3," Bull. Amer. Math. Soc., 79 (1973), 1006-1010.

[12] W. W. Boone, W. Haken, and V. Poénaru, "On recursively unsolvable problems in topology and their classification," *Contributions to Mathematical Logic*, North-Holland Pub. Co., Amsterdam (1968), 37-74.

[13] C. E. Burgess and J. W. Cannon, "Embeddings of surfaces in E^3," Rocky Mountain J. Math 1 (1971), 260-344.

[14] Sylvain E. Cappell and Julius L. Shaneson, "Invariants of 3-manifolds," Bull. Amer. Math. Soc. 81 (1975), 559-562.

[15] R. Crowell, "The group G'/G'' of a knot group G," Duke Math. J. 30 (1963), 349-354.

[16] R. Crowell and R. H. Fox, *Introduction to Knot Theory*, Ginn and Co. (1963).

[17] M. Dehn, "Über die Topologie des dreidimensionalen Raumes," Math. Ann. 69 (1910), 137-168.

[18] ———, "Die Gruppe der Abbildungsklassen," Acta. Math. 69 (1938), 135-206.

[19] D. B. A. Epstein, "Finite presentations of groups and 3-manifolds," Quart. J. Math. Oxford, 12 (1961), 205-212.

[20] ———, "Projective planes in 3-manifolds," Proc. London Math. Soc. (3) 11 (1961), 469-484.

[21] ———, "Curves on 2-manifolds and isotopies," Acta Math. 115 (1966), 83-107.

[22] B. Evans, "Boundary respecting maps of 3-manifolds," Pacific J. Math. 42 (1972), 639-655.

[23] B. Evans and L. Moser, "Solvable fundamental groups of compact 3-manifolds," Trans. Amer. Math. Soc. (1972), 189-210.

[24] C. D. Feustel, "A generalization of Kneser's conjecture," Pacific J. Math., 46 (1973), 123-130.

[25] C. D. Feustel and R. J. Gregorac, "On realizing HNN groups in 3-manifolds," Pacific J. Math. 46 (1973), 381-387.

[26] R. H. Fox, "On the complementary domains of a certain pair of inequivalent knots," Ned. Akad. Wetensch. Indag. Math. 14 (1952), 37-40.

[27] ———, "Construction of simply connected 3-manifolds," *Topology of 3-manifolds*, Prentice Hall (1962), 213-216.

[28] D. E. Galewski, J. G. Hollingsworth, and D. R. McMillan, Jr., "On the fundamental group and homotopy type of open 3-manifolds," Gen. Topology and its Applications 2 (1972), 299-313.

[29] L. C. Glaser, *Geometrical Combinatorial Topology* (2 vols.), Van Nostrand Reinhold Co. (1970, 1973).

[30] V. K. A. M. Gugenheim, "Piecewise linear isotopy," Proc. London Math. Soc. 31 (1953), 29-53.

[31] Wolfgang Haken, "Theorie der Normal Flächen," Acta. Math. 105 (1961), 245-375.

[32] ———, "Some results on surfaces in 3-manifolds," Studies in Modern Topology, Math. Assoc. Amer., distributed by Prentice Hall (1968), 39-98.

[33] ———, "Algebraically trivial decompositions of homotopy 3-spheres," Ill. J. Math. 12 (1968), 133-170.

[34] Wolfgang Haken, "On homotopy 3-spheres," Ill. J. Math. 10 (1969), 159-178.

[35] M. Hall, Jr., "Coset representations in a free group," Trans. Amer. Math. Soc. 67 (1949), 421-432.

[36] W. Heil, "On P^2-irreducible 3-manifolds," Bull. Amer. Math. Soc. 75 (1963), 772-775.

[37] ———, "Some finitely presented non 3-manifold groups," preprint.

[38] John Hempel, "Residual finiteness of Surface groups," Proc. Amer. Math. Soc., 32 (1972), 323.

[39] ———, "One sided incompressible surfaces in 3-manifolds," Lecture Notes in Math. No. 438 Springer-Verlag (1974), 251-258.

[40] ———, "Free cyclic actions on $S^1 \times S^1 \times S^1$," Proc. Amer. Math. Soc. 48 (1975), 221-227.

[41] John Hempel and W. Jaco, "3-manifolds which fiber over a surface," Amer. J. Math. XCIV (1972), 189-205.

[42] ———, "Fundamental groups of 3-manifolds which are extensions," Ann. of Math. 95 (1972), 86-98.

[43] I. N. Herstein, *Noncommutative Rings*, Carus Mathematical Monographs No. 15 (1968).

[44] Hugh M. Hilden, "Every closed orientable 3-manifold is a 3-fold branched covering space of S^3," Bull. Amer. Math. Soc. 80 (1974), 1243-1244.

[45] K. Hirsch, "On infinite solvable groups," Proc. London Math. Soc. (2) 49 (1946), 184-194.

[46] F. Hirzebruch, W. P. Neumann, and S. S. Koh, *Differentiable Manifolds and Quadratic Forms*, M. Dekker (1971).

[47] J. F. P. Hudson, *Piecewise linear topology*, W. A. Benjamin, Inc. (1969).

[48] William Jaco, "Heegaard splittings and splitting homomorphisms," Trans. Amer. Math. Soc. 144 (1969), 365-379.

[49] ———, "Three-manifolds with fundamental group a free product," Bull. Amer. Math. Soc. 75 (1969), 972-977.

[50] ———, "Surfaces embedded in $M^2 \times S^1$," Canadian J. Math. XXII (1970), 553-568.

[51] ———, "Finitely presented subgroups of three-manifold groups," Invent. Math. 13 (1971), 335-346.

[52] ———, "The structure of 3-manifold groups," mimeo notes, Princeton University (1972).

[53] I. Johansson, "Uber singuläre Elementarflächen und das Dehnsche Lemma," Math. Ann. 110 (1935), 312-330.

[54] A. Karrass and D. Solitar, "On finitely generated subgroups of a free group," Proc. Amer. Math. Soc. 62 (1969), 209-213.

[55] R. Kirby and L. C. Siebenmann, "On the triangulation of manifolds and the hauptvermulung," Bull. Amer. Math. Soc., 75 (1969), 742-749.

[56] J. M. Kister and D. R. McMillan, Jr., "Locally euclidean factors of E^4 which cannot be embedded in E^3," Ann. of Math. 76 (1962), 541-546.

[57] H. Kneser, "Geschlossene Flächen in dreidimensionalen Mannigfaltigkeiten," Jahresbericht der Deut. Math. Verein., 38 (1929), 248-260.

[58] R. Lee, "Semicharacteristic classes," Topology 12 (1973), 183-199.

[59] W. B. R. Lickorish, "A representation of orientable combinatorial 3-manifolds," Ann. of Math. 76 (1962), 531-540.

[60] ———, "A finite set of generators for the homeotopy group of a 2-manifold," Proc. Camb. Phil. Soc. 60 (1964), 769-778.

[61] G. R. Livesay, "Fixed point free involutions on the 3-sphere," Ann. of Math., 72 (1960), 603-611.

[62] Wilhelm Magnus, "Residually finite groups," Bull. Amer. Math. Soc. 75 (1969), 305-315.

[63] Wilhelm Magnus, A. Karrass, and D. Solitar, *Combinatorial group theory*, John Wiley and Sons (1966).

[64] A. I. Mal'cev, "On isomorphic matrix representations of infinite groups," Math. Sb. 8 (1940), 405-421.

[65] A. A. Markov, "Insolubility of the problem of homeomorphy," Proc. Inter. Congress Math. 1958, Cambridge University Press, 300-306.

[66] William S. Massey, *Algebraic Topology*, Harcourt, Brace, and World (1967).

[67] J. Mayland, "On residually finite knot groups," Trans. Amer. Math. Soc. 168 (1972), 221-232.

[68] D. R. McMillan, Jr., "Some contractible open 3-manifolds," Trans. Amer. Math. Soc. 102 (1962), 373-382.

[69] J. Milnor, "Groups which act on S^n without fixed points," Amer. J. Math. 79 (1957), 623-630.

[70] ———, "A unique factorization theorem for 3-manifolds," Amer. J. Math. 84 (1962), 1-7.

[71] E. E. Moise, "Affine structures in 3-manifolds V. the triangulation theorem and Hauptvermutung," Ann. of Math. 55 (1952), 96-114.

[72] J. M. Montesinos, "A representation of closed, orientable 3-manifolds as 3-fold branched covers of S^3," Bull. Amer. Math. Soc. 80 (1974), 845-846.

[73] J. R. Munkres, "Obstructions to smoothing piecewise-differentiable homeomorphisms," Ann. of Math., 72 (1960), 521-554.

[74] K. Murasugi, "On a certain subgroup of the group of an alternating link," Amer. J. Math., 85 (1963), 544-550.

[75] L. P. Neuwirth, *Knot Groups*, Annals of Math Studies No. 56, Princeton University Press (1965).

[76] M. H. A. Newman, "On the foundations of combinatorial analysis situs," Proc. Royal Acad. Amsterdam, 29 (1926), 611-641.

[77] ______, "On the superposition of n-dimensional manifolds," J. London Math. Soc., 2 (1927), 56-64.

[78] J. Nielsen, "Die Struktur periodischer Transformationen von Flächen," Math.-fys. Meddelelser Kgl. Danske Vidensk. Selsk XV (1937), 1-75.

[79] ______, "Abbildungsklassen endlicher Ordnung," Acta Math. 75 (1942), 23-115.

[80] C. D. Papakyriakopoulos, "On Dehn's lemma and the asphericity of knots," Ann. of Math. 66 (1957), 1-26.

[81] ______, "On Solid Tori," Proc. London Math. Soc. VII (1957), 281-299.

[82] ______, "A reduction of the Poincaré conjuncture to group-theoretic conjectures," Ann. of Math. 77 (1963), 250-305.

[83] H. Poincaré, "Second complément à l'Analysis Sitis," Proc. London Math. Soc., 32 (1900), 277-308.

[84] ______, "Cinquième complément à l'Analysis Sitis," Rend. Circ. Math. Palermo, 18 (1904), 45-110.

[85] E. S. Rapaport, "On the commutator subgroup of a knot group," Ann. of Math. 71 (1960), 157-162.

[86] P. M. Rice, "Free actions of Z_4 on S^3," Duke Math. J. 36 (1969), 749-751.

[87] G. Ritter, "Free actions of Z_8 on S^3," Trans. Amer. Math. Soc., 181 (1973), 195-212.

[88] C. P. Rourke and B. J. Sanderson, *Introduction to Piecewise-Linear Topology*, Ergeb. der Math. u. ihrer Grenz. 69 Springer (1972).

[89] G. P. Scott, "Finitely generated 3-manifold groups are finitely presented," J. London Math. Soc. (2) 6 (1973), 437-440.

[90] ______, "Compact submanifolds of 3-manifolds," J. London Math. Soc. (2) 7 (1973), 246-250.

[91] H. Seifert, "Topology of 3-dimensional fibered spaces," Acta. Math. 60 (1933), 147-288.

[92] A. Shapiro and J. H. C. Whitehead, "A proof and extension of Dehn's lemma," Bull. Amer. Math. Soc. 64 (1958), 174-178.

[93] L. C. Siebenmann, "Are nontriangulable manifolds triangulable?," *Topology of Manifolds*, Markham Publishing Co. 1970, 77-85.

[94] Jonathon Simon, "Some classes of knots with property P," *Topology of Manifolds, Proceedings of the University of Georgia Topology of Manifold Institute*, Markham Publishing Co. (1970), 195-199.

[95] John Stallings, "On the loop theorem," Ann. of Math. 72 (1960), 12-19.

[96] ____________, "On fibering certain 3-manifolds," *Topology of 3-manifolds*, Prentice-Hall (1962), 95-100.

[97] ____________, "A topological proof of Grushko's theorem on free products," Math. Zeit. 90 (1965), 1-8.

[98] ____________, "How not to prove the Poincaré conjecture," Ann. of Math. Study No. 60 (1966), 83-88.

[99] ____________, "On torsion free groups with infinitely many ends," Ann. of Math. 88 (1968), 312-334.

[100] P. Stebe, "Residual finiteness of a class of knot groups," Comm. Pure and Appl. Math., XXI (1968), 563-583.

[101] R. Swan, "Groups of cohomological dimension one," J. Algebra 12 (1969), 588-610.

[102] C. Thomas, "Nilpotent groups and compact 3-manifolds," Proc. Camb. Phil. Soc. 64 (1968), 303-306.

[103] F. Waldhausen, "Gruppen mit Zentrum und 3-dimensionale Mannigfaltigkeiten," Topology 6 (1967), 505-517.

[104] ____________, "Eine Verallgemeinerung des Schleifensatzes," Topology 6 (1967), 501-504.

[105] ____________, "On irreducible 3-manifolds which are sufficiently large," Ann. of Math. 87 (1968), 56-88.

[106] ____________, "Heegaard-Zerlegungen der 3-sphäre," Topology 7 (1968), 195-203.

[107] A. H. Wallace, "Modifications and cobounding manifolds," Canadian J. Math. 12 (1960), 503-528.

[108] J. H. C. Whitehead, "A certain open manifold whose group is unity," Quart. J. Math. (2), 6 (1935), 268-279.

[109] ______________, "Simplicial spaces, nuclei, and m-groups," Proc. London Math. Soc. 45 (1939), 243-327.

[110] J. H. C. Whitehead, "On 2-spheres in 3-manifolds," Bull. Amer. Math. Soc. 64 (1958), 161-166.

[111] ——————, "On finite cocycles and the sphere theorem," Colloquium Mathematicum 6 (1958), 271-281.

[112] ——————, "Manifolds with transverse fields in euclidean space," Ann. of Math. 73 (1961), 154-212.

[113] A. Wright, "Mappings from 3-manifolds onto 3-manifolds," Trans. Amer. Math. Soc. 167 (1972), 479-495.

[114] E. C. Zeeman, Seminar on Combinatorial Topology (mimeo notes), I. H. E. S., Paris, and Univ. of Warwick, (1963-1966).

[115] H. Zieschang, "On extensions of fundamental groups of surfaces and related groups," Bull. Amer. Math. Soc. 77 (1971), 1116-1119.

[116] ——————, "Addendem to: On extensions of fundamental groups of surfaces and related groups," Bull. Amer. Math. Soc. 80 (1974), 366-367.

INDEX

SYMBOLS AND NOTATION

B^n	the n-cell
∂M	boundary of M
$\beta_i(X)$	i^{th} Betti no. = rank $H_i(X)$
$\chi(X)$	Euler characteristic
$[g_1, g_2]$	the commutator $g_1 g_2 g_1^{-1} g_2^{-1}$
G'	the commutator subgroup of G
$[G, G]$	the commutator subgroup of G
$G_1 * G_2$	the free product of groups G_1, G_2
$[G:H]$	the index of a subgroup H in G
$\mathcal{G}$	the class of fundamental groups of compact, prime 3-manifolds (p. 170)
$H < G$	H is a subgroup of G
$HNN(B, C_0, C_1, \phi)$	(p. 173)
K', K''	the first, second barycentric subdivision of a complex K
$K^{(q)}$	the q-skeleton of a complex K
$K(G, 1)$	an aspherical complex with fundamental group G (p. 75)
$\ker \phi$	the kernel of a map ϕ
$lk(\sigma, K)$	the link of a simplex σ in a complex K (p. 5)
$L_{p,q}$	lens space of type (p, q) (p. 21)
$\hat{M}$	the manifold obtained by capping off the 2-spheres in ∂M (p. 25)
$M_1 \# M_2$	the connected sum of M_1 and M_2 (p. 24)
$MC(T)$	the mapping class group of T (p. 162)

P^n	real projective n-space
$\mathcal{P}(M)$	the Poincaré associate of M (p. 88)
$p(G)$	$\inf\{\#A : A$ generates $G\}$
R^n	real n-space
S^n	the n-sphere
$S(f)$	$Cl\{x : \#(f^{-1}(f(x))) > 1\}$ (p. 9)
$\Sigma(f)$	$f(S(f))$
$st(\sigma, k)$	the star of a simplex σ in a complex k (p. 4)
$\theta_i(G)$	$\cup\{H < G : [G:H] \leq i\}$ (p. 177)
$\mathfrak{W}$	the class of P^2-irreducible, sufficiently large 3-manifolds (p. 136)
$\omega(X)$	the subgroup of $\pi_1(X)$ of orientable loops (p. 22)
$\langle X:R \rangle$	the group presentation with generators X and relators R
Z	the infinite cyclic group
Z_n	the cyclic group of order n

ANNALS OF MATHEMATICS STUDIES

Edited by Wu-chung Hsiang, John Milnor, and Elias M. Stein

86. 3-Manifolds, *by* JOHN HEMPEL
85. Entire Holomorphic Mappings in One and Several Complex Variables, *by* PHILLIP A. GRIFFITHS
84. Knots, Groups, and 3-Manifolds, *edited by* L. P. NEUWIRTH
83. Automorphic Forms on Adele Groups, *by* STEPHEN S. GELBART
82. Braids, Links, and Mapping Class Groups, *by* JOAN S. BIRMAN
81. The Discrete Series of GL_n over a Finite Field, *by* GEORGE LUSZTIG
80. Smoothings of Piecewise Linear Manifolds, *by* M. W. HIRSCH *and* B. MAZUR
79. Discontinuous Groups and Riemann Surfaces, *edited by* LEON GREENBERG
78. Strong Rigidity of Locally Symmetric Spaces, *by* G. D. MOSTOW
77. Stable and Random Motions in Dynamical Systems: With Special Emphasis on Celestial Mechanics, *by* JÜRGEN MOSER
76. Characteristic Classes, *by* JOHN W. MILNOR *and* JAMES D. STASHEFF
75. The Neumann Problem for the Cauchy-Riemann Complex, *by* G. B. FOLLAND *and* J. J. KOHN
74. Lectures on *p*-Adic *L*-Functions *by* KENKICHI IWASAWA
73. Lie Equations *by* ANTONIO KUMPERA *and* DONALD SPENCER
72. Introduction to Algebraic *K*-Theory, *by* JOHN MILNOR
71. Normal Two-Dimensional Singularities, *by* HENRY B. LAUFER
70. Prospects in Mathematics, *by* F. HIRZEBRUCH, LARS HÖRMANDER, JOHN MILNOR, JEAN-PIERRE SERRE, *and* I. M. SINGER
69. Symposium on Infinite Dimensional Topology, *edited by* R. D. ANDERSON
68. On Group-Theoretic Decision Problems and Their Classification, *by* CHARLES F. MILLER, III
67. Profinite Groups, Arithmetic, and Geometry, *by* STEPHEN S. SHATZ
66. Advances in the Theory of Riemann Surfaces, *edited by* L. V. AHLFORS, L. BERS, H. M. FARKAS, R. C. GUNNING, I. KRA, *and* H. E. RAUCH
65. Lectures on Boundary Theory for Markov Chains, *by* KAI LAI CHUNG
64. The Equidistribution Theory of Holomorphic Curves, *by* HUNG-HSI WU
63. Topics in Harmonic Analysis Related to the Littlewood-Paley Theory, *by* ELIAS M. STEIN
62. Generalized Feynman Amplitudes, *by* E. R. SPEER
61. Singular Points of Complex Hypersurfaces, *by* JOHN MILNOR
60. Topology Seminar, Wisconsin, 1965, *edited by* R. H. BING *and* R. J. BEAN
59. Lectures on Curves on an Algebraic Surface, *by* DAVID MUMFORD
58. Continuous Model Theory, *by* C. C. CHANG *and* H. J. KEISLER
56. Knot Groups, *by* L. P. NEUWIRTH
55. Degrees of Unsolvability, *by* G. E. SACKS
54. Elementary Differential Topology (Rev. edn., 1966), *by* J. R. MUNKRES
52. Advances in Game Theory, *edited by* M. DRESHER, L. SHAPLEY, *and* A. W. TUCKER
51. Morse Theory, *by* J. W. MILNOR
50. Cohomology Operations, *lectures by* N. E. STEENROD, *written and revised by* D.B.A. EPSTEIN
48. Lectures on Modular Forms, *by* R. C. GUNNING
47. Theory of Formal Systems, *by* R. SMULLYAN

Library of Congress Cataloging in Publication Data

Hempel, John, 1935-
3-manifolds.

(Annals of mathematics studies ; no. 86)
Includes bibliographical references and index.
1. Manifolds (Mathematics) I. Title. II. Series.
QA613.H45 1976 516'.07 76-3027
ISBN 0-691-08178-6
ISBN 0-691-08183-2 pbk.